COMPANION PLANTS

Companion Plants
&
How to Use Them

❧ *Helen Philbrick*
❧ *Richard B. Gregg*
❧ INTRODUCTION: *Dr. agr. H. H. Koepf*

Published by
Biodynamic Farming and Gardening Association, Inc.

Printed by
Partners Press, Inc.
OAKS, PENNSYLVANIA

ISBN: 978-0-938250-05-0
Library of Congress Catalogue Number: 65-19128
Printed in the United States of America

Ninth printing, 1977; Seventeenth printing, 1991

Note:

In 1943, a first attempt was made to collect information on plants that did well as companions, and those that seemed to evidence antagonisms. The work was done by Richard Gregg, who used the garden of Evelyn Speiden to conduct certain experiments. A pamphlet resulted, published by the Biodynamic Farming & Gardening Association. This book includes that material. It also includes the work of the late Dr. E. E. Pfeiffer, in the fields of sensitive crystallization and chromatography, ably recorded by his assistant, Erica Sabarth, to whom special thanks are due.

Contents ℥ ℥ ℥

THE STRAWBERRY GROWS UNDERNEATH THE NETTLE, AND WHOLESOME BERRIES THRIVE AND RIPEN BEST, NEIGHBORED BY FRUIT OF BASER QUALITY...

King Henry V, ACT I, SCENE I

Preface ❧ ❧ ❧
Companion &
Protective Plants

Over a long period of years, people have observed from
time-to-time that certain plants seem to affect, favorably
or unfavorably, certain other plants which may be growing
near them. The observers were sometimes peasants or farmers, sometimes gardeners, sometimes scientists such as
botanists. The significance of this type of phenomena has
been apparently appreciated most clearly by those who were
interested in the biodynamic methods of farming and gardening. At any rate, they have made the largest assemblage of
instances of this sort.

A knowledge of these kinds of relationships between plants
means that by planting in accordance therewith one can
secure more than the usual control of results. The quality of
food and flowers is improved, the losses from insect pests and
plant diseases are reduced, the time and money needed to
ward off such attacks are lessened; because of the increased

vigor of the plants, the risks of losses from drought and frost are made smaller, the money profit and satisfaction and pleasure of work are enhanced. Furthermore, the observation of these relationships stimulates imagination and sensitiveness of observation to other living relationships and thereby opens new doors to further understanding of the world of nature. One can also avoid the failures that result from putting plant opponents near one another.

There is now a goodly number of people who could use such information on their farms and gardens and who would like to experiment with it. We have, therefore, tried to compile and arrange it for their benefit. We present it here, hoping that readers will send in other instances they have observed, or of which they have been reliably informed.

RICHARD B. GREGG

Introduction ❧ ❧ ❧

The Biodynamic Method of farming and gardening is based on the study of (and a steadily expanding knowledge about) the mutual influences of living organisms. Emphasis is placed on practical observation of the effects exerted on plants and animals by their earthly and cosmic environment. Thus, this method supplements the approach now commonly in use in farming and gardening. This latter approach is essentially chemical and physical, since one isolates the building bricks of life and the substances which control the metabolism of plants and animals. These are then applied in order to increase the yields or to eradicate undesirable pests and weeds.

In the early days of the development of this physical-chemical analytical path, there were only a few inorganic elements believed to be worth applying to plants. But the list grew longer and, as time went on, minor elements joined the major nutrients. Recently, not only inorganic but also organic compounds were found to be active. Among these are root excretions, organic substances present in leaves, stubble, roots, etc. Thus, the picture becomes more and more complicated. The search for the biological principles that steer this tangled complex of effects and counter-effects is an important one. The plant itself is a mighty factor in this. It has specific influences on the other plants and on the micro-life of the soil. This is a field of knowledge to which both practical experience and scientific research still have much to contribute. It may be expected that such studies will provide increasing insight into the many interrelationships amongst the plants and animals forming the community of a particular habitat. A quotation from Goethe expresses what we are speaking of here, "Nothing happens in living Nature that is not in relation to the whole." This may be termed one of the leading ideas of the biodynamic concept.

Plants growing together within the relatively small space of a garden or field may interact because one tall-growing species gives shade, thus temporarily helping or suppressing another species. Again, plants may interact simply in competing for water, or nutrients. Much is known about this. One species may benefit the other because it forms a deep and luxuriant root system, thus loosening the ground, or, because it enriches the soil with organic substances and nutrients. There may also be the effects of excretions, odors, insect repelling or attracting substances, biotic compounds, and so on. These may directly influence the growth of other plants, or alter the population of micro organisms that live in the soil, or be effective in the crowded world of animals crawling and flying in and around the roots, leaves and blossoms.

All of these factors are known to have practical significance in growing high yields of tasty produce, with high keeping-quality. An increasing use of the beneficial effects one species exerts on another will help more and more toward the desirable end of first decreasing, then dropping the application of strong poisons for pest and disease control. This opinion is backed by a steadily growing body of practical observations. More than this can happen, however, when we are working in this way. We may well discover a steadily flowing source of pleasure, and be stimulated to admiration for the wonders our everyday environment can offer.

It was a truly worthwhile task, which Richard B. Gregg performed in the early forties when he gathered the information, then obtainable, about companion plants. This material was printed in *Biodynamics*, III, 1, 1943, and later put out as a pamphlet, together with an article by Evelyn Speiden Gregg, about the value of herbs in the garden. This small pamphlet has since then reached many people. Mrs. Helen Philbrick, working together with the Greggs, has now prepared a new and much expanded edition. New material has been added; the whole has been arranged alphabetically; and an index lists the common and scientific names of the species mentioned. In a modest sense, the present volume may also be considered to be the fruit of teamwork, since a number of active members of the Biodynamic Farming and Gardening Association have contributed to the manuscript.

This little book is definitely expected to stimulate teamwork in a wider sense. It is not a book to be read once, but one to keep at hand – for daily work in the garden and in the field. Those who work along these lines are asked to share their knowledge with the author – by confirming what seems to work, by disapproval of the measures that are questionable, or, do not work at their place, and by adding new and old information not included here. Thus, one may hope that within a reasonable span of time, an improved and still more enlarged edition dealing with this stimulating subject can be prepared.

When we look at plants as forms, as metabolic systems, or as systems that react in a specific way to environmental influences, we shall find that each species represents a particular type. The type expresses itself in manifold morphological, physiological, and biological properties. Taken as a whole it cannot be confounded with other types. The details are related to the whole. The type is an organizing principle that manifests the fact that a particular species is not just an accidental conglomerate of properties. E. E. Pfeiffer's method of sensitive crystallization provides a new way to demonstrate visibly some of the workings of this organizing principle.

Space does not permit more than a brief, generalized sketch of the crystallization method here, further details will be found in the chapter on its use in testing plant relationships. When a copper chloride solution crystallizes on a glass plate, its tiny crystals form an irregular pattern. If extracts from biological materials are added to the crystallizing salt, then a coordinated pattern is produced. Typical patterns have been established for many plant species and plant organs, as a result of the influence of environmental factors and daily rhythms. The crystallization pictures, as they are called, gave many insights on how matter is organized to bring about a particular type or species. Inevitably, this method has also been employed to study the relationships between different plant species. Thus, the ecological approach to an understanding of plant companionship was supplemented by a testing method. Some of the material gathered by Doctor Pfeiffer and his co-workers is reported by E. Sabarth in a contribution to this book. Further experiments and research, are of course, needed. They are a part of the work that still waits to be completed.

Emphasis must be placed on the fact that the material presented here is based on very different levels of experimental evidence. Those who prepared this edition realize this. Much of the information given has been gathered from biodynamic workers the world over. It is based on their own observations, or has been taken from older literature. Some of the advice has been confirmed, by actual use of it, over a prolonged period of time, although a true explanation may still be lacking. Other statements may be just occasional observations made here or there, and handed over from one writer, or lecturer, to another. Some of the information is taken from scientific literature, which at this time is rapidly increasing its coverage of this field. This kind of information does not yet have a definite character. Many of the findings derive from experiments carried out in the test tube and in the greenhouse. They have still to prove their efficacy under field conditions. It is hoped that in the near future, the time will come when real knowledge will be screened out of casual observations or preliminary suggestions. At this moment, the authors feel that spreading information and stimulation are more important than too much sifting.

This book has been compiled, chiefly, for practical farmers and gardeners. It is our hope, that it will promote further progress in a field which deserves constant attention, by those working with the Biodynamic Method, and by others, as well.

H. H. KOEPF
SPRING VALLEY, N.Y.

The Companionate Alphabet 🌿 🌿 🌿

ABSINTHIUM *(Artemisia absinthium)* Other names commonly used for this plant are Vermouth, or Wormwood.

(See WORMWOOD*)*

AFFORESTATION Strongly vital plants have the power to prepare unfavorable soil so that it will support the growth of trees. Three of these plants which are especially strong and vital in themselves are blackberries, Stinging Nettle, and couch (or quack) grass.

ALDER *(Alnus tenuifolia)* One of many Alders and closely related to the European Black Alder *(A. glutinosa).*
The chief role of alders is to anchor streams to their beds and to help drain wet soils.

ALFALFA *(Medicago sativa)* Some farmers know that where Dandelions make good growth, alfalfa will grow well too.

(See DANDELION*)*

Where alfalfa grows in pastures it gives protective shelter with its deep roots for shallow-rooting grasses. It hinders evaporation and remains green longer. It thus helps to keep its companion plants alive for longer periods.

(See LEGUMES*)*

ANISE *(Pimpinella anisum)* Anise seed will germinate and grow better if it is sown with Coriander seed. Formation of the seed itself in the Anise plant is strengthened by the presence of Coriander. *(See* WORMWOOD*)*

ANTS *(Formicidae)* Ants are repelled by Pennyroyal and Spearmint. Scatter these on pantry shelves to prevent infestations by ants, or to drive them away once they have come.

Mints planted by the front and back door of a house have prevented ants from entering. Tansy planted around a house, or placed on shelves inside, is also a repellent to ants.

APHIS *(Aphididae)* Nasturtiums planted near broccoli keep the latter free from aphis. *(Research suggested)*

APPLE TREES *(Malus)* Apple trees infested with scab have been helped by Chives growing near their roots. Another remedy for scab is composted pigeon manure and Equisetum, *(horsetail)*, tea.

(See chives, horse-radish, AND *horsetail teas under* HERBS*)*

If apple trees are surrounded by a few nasturtium plants, woolly aphis are repelled.

Scientists have shown that there are root excretions from grass which have suppressed the growth of the young, peripheral root tips of apple trees. Since these roots are the most active part of the apple tree's root system, the inhibitory effect of the grass roots can be serious.

In 1947, Bordukova studied the effect of apple trees on various plants, and found that the nearness of apple trees caused potatoes to be more susceptible to Phytophthora blight.

Ripening apples give off ethylene gas in infinitesimal amounts. This gas may inhibit the height of growth of neighboring plants, and cause flowers and fruits of neighboring plants to mature early. Dr. Pfeiffer used to caution against storing carrots in a root cellar near apples because the carrots would take on a bitter flavor. Traditionally, New Englanders know that apples and potatoes should not be stored in the same root cellar, because apples will lose their flavor and the potatoes will have an off-flavor and not keep well.

APRICOTS *(Prunus armeniaca)* Experiments reported in 1905 tell of an inhibitory effect of root excretions of oats on the growth of young apricot trees. Root excretions of potatoes and tomatoes were also tested but had less effect. Mustard and Rape affected the apricot trees still less, while beans and clover did not suppress growth at all.

ASPARAGUS *(Asparagus officinalis)* Asparagus is aided directly by tomatoes which in turn are aided by the asparagus. This "fact of life" in the perennial bed is of considerable help to the home gardener who often finds weeding the asparagus bed several times during the season an unpleasant chore. After the asparagus harvest in late spring, the soil is ready to receive the more tender tomato plants. Cultivation of the latter during their long growing season, up until frost, insures that there will be no weeds overgrowing the asparagus bed. But this is only one angle of their mutual benefit. There is a relationship between the two growing plants which insures their compatibility.

Asparagus also benefits by the presence of parsley which fits nicely into the pattern with tomato plants.

One of the few insects troublesome to asparagus is the asparagus beetle *(Crioceris asparagi)*, which may be controlled by letting the family hens run in the asparagus bed; but some way should be found to protect the tomatoes because hens are fond of the fruit even while it is green and immature.

A substance named asparagin has been isolated which has a good influence on tomato plants, helping to control some of the soil pests that affect them.

AUXINS   *(See* GROWTH REGULATORS*)*

BASIL *(Ocimum basilicum)* After stating over and over again that aromatic herbs and summer flowering plants are beneficial to the vegetable garden because they enliven and stimulate the sometimes heavy and monotonous quality of some vegetable plants, warning must be given that Sweet Basil does not like Rue. Perhaps this is a disharmony that the observant gardener should recognize from the inherent nature of each plant: Basil being one of the sweetest and Rue being of all plants surely the most bitter.

BEANS *(Phaseolus* AND *Vicia)* Since beans are probably one of our most popular and useful vegetables, and one most often found in the small garden, many plant combinations both good and bad have been observed relating to them: Beans thrive best with carrots and cauliflower interplanted. Beans are aided by carrots. Beans and beets also do well together. They also help cucumbers and cabbages. Beans planted with leek and celeriac in moderate quantity are a good combination, but if the beans are planted too thickly between leeks and celeriac, all three become stunted. In general beans are inhibited by onions, garlic and shallots.   *(See* SAVORY*)*
In a large planting of corn, broad beans are a good companion crop. Even in the small backyard garden corn and pole beans grow well together because the pole beans can climb up the corn stalks. Also in large scale

planting of broad beans and field beans, oats are a good plant affinity.

For a border around a small garden, pole beans are excellent. They can also be used to advantage as a border to protect a block of corn rows from too much wind. The beans add nitrogen to the soil, which is used by the corn.

Corn rows can be used as a windbreak for bush beans which in turn add nitrogen to the soil to be used by the corn.

BEANS *and* **POTATOES** Mr. George Corrin, Consultant of the Biodynamic Agricultural Association of Great Britain, wrote his observation, published in *Star and Furrow*, that broad beans *(vetch family)* and early potatoes always do well together. English gardeners have had plenty of seasons in which to experiment with this combination, for the first literary reference to combining broad beans with potatoes was written in 1779 by William Speechly, who recommended making rows three feet apart, with potatoes in the rows spaced 18 inches apart. Between the potato plants he transplanted broad bean plants, and mentioned the fact that the beans, being transplanted earlier, were ahead of both potatoes and weeds in growth. *(Loudon's Encyclopedia)* American gardeners for many years have made it a practice to plant green beans, bush or dwarf, in alternating rows with potatoes. Some years the insect control has been better than other years, but in general it has been found that the potatoes repel the Mexican bean beetles while the beans keep the Colorado potato beetle down enough so it can be controlled by hand picking.

BUSH, DWARF, *and* **GREEN BEANS,** also known as **SNAP, STRING, WAX,** *or* **BUTTER BEANS** *(Phaseolus vulgaris)* Bush beans aid celery if planted in ratio of one to six plants. That is, one bean plant to six of celery.

Bush beans and cucumbers are mutually helpful if beans are planted in a border around the cucumbers. Dwarf beans grow better with cabbages than with beans alone.

Keep beans and Fennel far apart. Their dislike is mutual.

Bush beans and strawberries like each other. When the earliest biodynamic publications were printed here in the 1930s, it was recorded that beans and strawberries were mutually beneficial. The writings did not state exactly which kind of beans. Several experiments were set up and it became evident that bush beans in the strawberry bed were measurably advanced over bush beans in the control plots. The strawberries also benefited noticeably from their proximity to the bush beans.

LIMA BEANS *(Phaseolus limensis)* Lima beans grow well near locust trees. Lima beans do not exercise the same control over the Colorado potato beetle, as observed by several biodynamic gardeners in one locality.

POLE BEANS Pole beans, like other beans, are suppressed by onions. It has been observed that beets and kohlrabi both grow wretchedly near pole beans. Radishes, on the other hand, do well near them.

BEECH *(Fagus)* Beech trees and ferns are antagonistic. Scilla grows exceptionally well under beech trees.

BEETS *(Beta vulgaris)* Beets like to grow near dwarf beans, onions, or kohlrabi. They grow poorly near pole beans, and they are harmed by Charlock and field mustard. Lettuce and cabbage like beets.

BIODYNAMIC VEGETABLE GARDEN A vegetable garden managed according to the biodynamic method follows certain definite principles of plant symbiosis such as the use of a well-planned crop rotation, companionate plantings, juxtaposition of deep rooted plants with shallow rooted plants, and the generous use of summer flowering shrubs and aromatic herbs all through the garden whether it be large or small. The observance of beneficial crop rotations and the careful use of companion plants, both in space (in one single year) and in time (from one year to another) make it possible to get maximum benefits from limited garden space with a minimum of labor without sacrificing the health-giving quality of the produce. Another most important factor of the biodynamic garden is that attention to all of these details insures that the garden soil is improved each year. Practice of the Biodynamic Method cares for the soil and leaves it in improved biological condition by good cultural practices, especially in the return of soil nutrients each year through the use of biodynamic compost.

BIRCH *(Betula)* It is good to have birch trees around the outskirts of compost and manure piles where they always encourage good fermentation. This is possibly caused by substances excreted by the birch roots.

If birch leaves are added to the compost heap, they will help exhausted soils. Soil taken from underneath birch trees or the soil surrounding their roots may be used to heal the ground where diseased plants have been removed. Such "birch soil" will also restore barren soil where pine or fir trees have discouraged fertility.

Before the widespread use of power machinery it used to be a common sight to see an American farmer "brushing," a harrowed field with several birch trees lashed together and dragged behind a horse or tractor. This practice always resulted in a fine seed bed.

BIRDS Birds are unequalled around the garden as controllers of insect pests, and they may be attracted by many different methods. Feeders and bird waterers will encourage them to frequent the garden and its surroundings. Some shrubs are especially attractive because of their edible seeds or fruits: hackberry, elderberry, mulberry, wild cherry, dogwood, Japanese barberry, viburnum and others. The bulletins of the United States Department of Agriculture and of the National Audubon Society on promotion of wild life will give names of other trees, shrubs and vines to attract bird life.

Evergreen trees and bushes, thorn bushes and honeysuckle vines are especially attractive to birds for building their nests. Bird houses may also be erected in garden areas to attract many different kinds of birds.

BLACK ALDER *(Ilex verticillata)* The Black Alder, through the action of its roots, helps drain wet soil. It also has root nodules which add nitrogen to the soil. It is the only nonleguminous shrub which has this power. Black Alder is not to be confused with Elder, or with the European Black Alder *(Alnus glutinosa),* which belongs to the birch family. Other names are Winterberry and Dogberry, and it can be identified by its bright red berries in late fall.

BLACKBERRIES *(Rubus allegheniensis)*   *(See* AFFORESTATION*)*

BLACK FLY (APHIS) *(Aphididae)* Black Fly is another name for the black aphis which attacks broad beans in particular. In addition to cutting off the tops of the bean plants and thus destroying the flies as well as the damaged tops, the Black Fly can be kept under control by the use of a fermented extract of nettle.   *(See* STINGING NETTLE*)*

Another principle of control, based upon COMPANION PLANTS, is mentioned under INTERCROPPING. A natural enemy of the Aphis is the Lady Beetle, which often increases and controls Black Fly, as long as this natural process is permitted, and is not upset by outside influences.

BLACKTHORN *(Prunus spinosa)* The leaves of this plant made into compost are of particular help to exhausted soils.

BORAGE *(Boraginaceae)* Since strawberries and Borage are mutually beneficial, if the Borage is limited to a small proportion, a few plants in the strawberry bed will do well, and help the strawberries. Some gardeners however, prefer to keep Borage outside the beds, because it sprawls and takes up a good deal of room. Honey bees show enthusiasm for Borage when it is in bloom.

BORDER PLANTS Borders of the following in small ratio have been found helpful to all vegetables: Dead Nettle *(Henbit),* which has no resemblance to Stinging Nettle *(Urtica dioica)*; Sainfoin, a pink flowered herb also called Esparsette; Valerian, often called Garden Heliotrope; also Hyssop and Lemon Balm, both of which will form a little hedge in two years. Yarrow can be used in an area that is used as a path because it does not object to footsteps. *(See* HYSSOP, LEMON BALM, SUMMER FLOWERING PLANTS AND SHRUBS, *and* VALERIAN.*)*

BROCCOLI *(Brassica oleraceae)* Broccoli in general follows the same rules of behavior as the other members of the cabbage family.

BROOM-RAPE *(Orobanche ramosa L.)* Broom-rape seeds germinate only in the vicinity of the roots of sunflower, flax, corn, soy and some other plants. This is a phenomenon explained by the fact that many plant roots, during their lifetime, excrete substances which stimulate the growth of other organisms. Broom-rape is the name given to various leafless herbs growing as parasites on the roots of other plants. There are some 90-odd species. Examples are Hemp broom-rape, Squawroot and Beechdrop *(Epifagus)*. The Pink Lady-slipper may be similarly conditioned. People who have extensive experience in growing wildflowers say that unless a certain root fungus is present in the soil, the Lady-slipper is impossible to transplant. Other observers familiar with areas in birch woods where formerly there were masses of Pink Lady-slippers report that the birch began to disappear and was succeeded by seedlings of oak and maple. When the birch died out, the Ground Pine increased and the Lady-slippers disappeared apparently because soil and neighboring plants were no longer compatible.

BUCKWHEAT *(Fagopyrum esculentum)* Buckwheat grows on the poorest soils and collects large quantities of calcium which it adds to the soil when it is plowed under. In addition, it chokes out weeds and will in time eradicate them. It also helps to loosen heavy soil. The calcium and the green matter plowed under will in time enrich the soil. Buckwheat is antagonistic to winter wheat.

BUTTERCUP FAMILY *(Ranunculus)* Members of the buttercup family secrete a substance in their roots which poisons the soil for clover, being hostile to the growth of nitrogen bacteria. Since this works even in so fine a dilution as 1-part to a billion, farmers find that clover disappears in a meadow where buttercups are increasing. Delphinium, peonies, monkshood, columbine and double buttercups belong to the same ranunculaceous family – a family strong and vital, but growing only for itself. These plants consume large amounts of humus; they are heavy feeders. Without plenty of good composted manure, do not expect other plants to grow well in their vicinity.

CABBAGE *(Brassicaceae)* When the cabbage family is mentioned it includes cauliflower, broccoli, brussel sprouts, kale and kohlrabi. A close study of the members of this family will reveal that each has been developed in a one-sided way. The cabbage has been developed to an enormous terminal bud, while its flowering process is subordinated. This one-sided evolution results in a life cycle that is easily upset and a plant which will decay rapidly under adverse conditions. The cabbage is greatly helped if it is surrounded by plants which have many blossoms, or which are strongly aromatic. Both of these characteristics help to balance the cabbage plant's weaknesses. Dip cabbage roots into a paste of cow manure, clay and Biodynamic Preparation 500 when transplanting. This is suggested as a help against club root. After transplanting, give cabbage seedlings a mulch of half-rotted compost. If this is given before the soil becomes dry, it will be a great help in case of drought.

Do not grow cabbages in the same place two years in succession because of the danger of club root.

Late cabbage and early potatoes do well together. Plant when the latter are first hoed up into hills. Cabbage is aided by Dill, Chamomile, Sage, Vermouth *(Wormwood)*, Rosemary, and in both quality and quantity, by members of the peppermint family.

Cabbage dislikes strawberries, but likes beets.

One of the chief pests of the cabbage family is the white cabbage butterfly, which may be repelled by the following plants growing near the cabbages: tomatoes, Sage, Rosemary, Hyssop, Thyme, Mint, Hemp, Wormwood and Southernwood *(Artemisia abrotanum)*.

In an experimental planting, tomatoes near cabbages kept the white cabbage butterfly away.

When cauliflower and Stinging Nettle occupy neighboring plots, if the cauliflower is planted too closely, it will increase the growth of the nettles.

CALCIUM Melon leaves are extremely rich in calcium. If such leaves are added to a compost heap the calcium content will be increased.

Buckwheat also accumulates calcium and, when composted or plowed under, enriches the soil with this element.

Scotch Broom belongs to the legume family and it too accumulates calcium. If it is kept in check, it thus aids the soil.

Another plant used to add calcium to the topsoil is the Lupine *(lupinus)*. This is especially applicable in the case of poor sandy soils.

(See MINERAL REQUIREMENTS OF PLANTS*)*

CAMOMILE (CHAMOMILE) *(Matricaria chamomilla)* This species, often called German Camomile, is one of the compost plants used in the biodynamic method. It is a specialist for lime. It also contains a growth hormone which particularly stimulates the growth of yeast. This growth hormone works most actively in Camomile extract (matricaria oil) in a dilution of 1:8 million. Under growing conditions Camomile in small quantities increases the essential oil content in the Peppermint plant. But as the proportion of Camomile plants increases, the amount of Peppermint oil diminishes. This information has been checked in many test plots under varying combinations. *(See* COMPOST PLANTS*)*.

Camomile in the wheat field, in a ratio of 1-Camomile plant to 100-wheat plants, helps the latter to grow heavy with fuller ears. But if Camomile is in thicker patches, it inhibits the growth of neighboring plants, and results in small, light seed. (The more commonly found Camomile called Mayweed, and others of the genus, *Anthemis,* often become obnoxious weeds; they are not cultivated as a rule.) Camomile also helps neighboring onions, but only if the ratio is 1-plant of Camomile to every 4-yards of onions. It is also good to grow with cabbage.

Camomile tea is effective against a number of plant diseases, especially when the plants to be treated are young. It can be used to control damp-off in greenhouses and cold frames. The tea is best prepared by soaking dried blossoms in cold water for a day or two.

Camomile tea is used for diarrhea or scours in calves. A steeped tea made from equal parts of Camomile, Chervil and Lemon Balm, applied in a compress, is excellent to cure hoof rot in animals. Keep the compress on 20 minutes and then dry off.

CARAWAY *(Carum carvi)* The long roots of the Caraway plant may be a substitute for subsoiling on heavy, wet land, since it leaves the soil crumbly. It is difficult to get Caraway to start growing, therefore it is best to start it with a companion crop of peas. Harvest the peas, harrow as though nothing were there — and the Caraway comes up later. It is well known that Caraway seed adds an aromatic flavor, and makes rather hard, heavy breads more digestible. It is also used in soups, cheese, and other dishes.

Caraway dislikes Fennel.

CARROTS *(Daucus carota)* To prepare heavy soil for a carrot crop, raise flax or soybeans for one year to loosen the soil and leave it friable.

H. Molisch made a study of allelopathy, which is the scientific name for plant companionships. He isolated an exudate from the roots of carrots which has a beneficial effect on the growth of peas. Carrots grow well with leaf lettuce and chives. Also with red radishes.

The Carrot Fly *(Psila rosae)* is a troublesome pest of carrots. It is the maggot or larva of this fly which attacks the rootlets of the young plant. It often pupates in the mature stored carrots. Various other plants have been found as repellents: onions and leeks and strong smelling herbs like Rosemary, Wormwood and Sage. As H. Thiess reports in *Lebendige Erde,* (May, 1963), Black Salsify *(Scorzonera hispanica)*, is effective in repelling the carrot fly. He plants a mixed crop of carrots and Black Salsify. The latter is often called the Oyster Plant, because of its flavor.

An English gardener recommends sowing carrot seed thinly, so as to avoid thinning, as he has observed that pulling the carrot seedlings seems to attract the carrot flies. He recommends sowing two parts leek and three parts carrot seed. The carrots will be ready to harvest early, which will give the leeks time to mature before fall. The presence of the leeks seems to repel the flies.

After carrots are harvested, do not store them in a cellar near apples, because the latter will make the carrots taste bitter.

(See CHIVES, WORMS IN GOATS*)*

CASTOR BEAN *(Ricinus communis)* The castor bean plant repels mosquitoes; but it is poisonous to livestock and to humans. It should be cultivated with caution. One or two seeds if eaten could be fatal to an adult.

CATNIP *(Nepeta cataria)*   *(See* HONEY BEES*)*

CAULIFLOWER *(Brassicaceae)* Cauliflower is known to grow better if there is celery in its neighborhood. It is reported that the white cabbage butterfly, *(Pieris rapae)*, is kept away from cabbage and cauliflower by nearby celery plants.

CELERIAC *(Apium graveolens rapaceum)* Celeriac is a variety of celery with an enlarged edible root, which should be better known and appreciated in America. It tastes like celery but is somewhat easier to raise. To prepare the soil for celeriac, it has been found that winter vetch is good. Sow the vetch early and let it grow into spring. When the weather is warm enough for sowing celeriac, cut the vetch and compost the green parts. Dig or plow the soil and put on composted manure lightly raked in and then sow the celeriac seed. This vegetable needs a rich and friable soil with plenty of potassium. The leek is also a potassium lover and is a good plant companion for celeriac. In such a combination the celeriac plants should be 12-15 inches apart and the leeks 8-12 inches apart in the next row, with the two rows alternating. Both need the whole season to mature and they will continue growth into the cool fall weather. Leeks may be left in the ground all winter but will rot or send up a seed stalk early the following spring. Scarlet Runner beans growing in rows next to celeriac resulted in a better celeriac crop than in a neighboring test plot where there were no beans.

The edible part of the celeriac plant, a large bulbous root, tastes very much like celery. It can be sliced or shredded for salads or soups. This bulbous root complements the tall thin stalk of the leek as the two plants grow together in the biodynamic garden bed. Celeriac is easier to raise than celery because it requires no blanching.

CELERY *(Apium graveolens)* Like celeriac, celery is benefited by leeks growing nearby with two rows of each alternating. Both plants also like composted pig manure. Both celery and leeks grow well in a trench. Tomatoes also are good neighbors to celery. Another good companion is the bush bean, which seems to help celery. For many years biodynamic farmers have followed the practice of planting one bush bean to every six celery plants.

In 1951, research was carried out in Germany, which included data on the protective effect of celery. The research was summarized, stating, "When cabbage grows in the vicinity of this plant, it is less affected by microorganisms." This is only one of many examples, where modern scientific data has upheld the "old-wives tales" of earlier generations.

CHARLOCK *(Brassica arvensis)* This is an annual weed of the mustard family. When Charlock seeds are present in the soil, they are awakened by the presence of oat seed.

CHERRY *(Prunus avium)* The roots of cherry suppress the growth of wheat. The strongest suppression was observed in the summer. Plant catnip under cherry trees to attract cats who will scare the birds away.

It was also shown that when there were cherry trees in the vicinity of potato plants, the latter were less resistant to Phytophthora (one of the blights).

CHERVIL *(Anthriscus cerefolium)* Chervil is described in the reference books as an herb of the easiest culture. We failed to find it so until we learned that it likes partial shade in summer. Now it covers the ground, and reseeds itself, with delicately cut low growing leaves with a faint aromatic aroma.

Chervil and radishes are mutually helpful plant companions and those who have made tests report that radishes growing nearest Chervil have a hotter taste.

It is worth noting that the old botanical name, Chaerophyllum, comes from two Greek words which together mean that the leaves bring gladness.

(See CHAMOMILE*)*

CHIVES *(Allium schoenoprasum)* Chives are so common that they are taken for granted in almost any garden. What often goes unnoticed is that the Chive plant is hardly ever attacked, either by disease, or by insects. Biodynamic gardeners who have always tried to make a very careful study of every plant in its relation to its environment have pondered this robust health of the Chive.

Experiments using Chives as companion plants have shown that they help apple trees to better health and that they prevent apple scab. Some orchardists have made Chive tea to spray against apple scab and against downy and powdery mildew on gooseberries and cucumbers.   *(Recipe for Chives tea under* HERBS*)*

There is a report of a very large bed of Chives – 120'x6'. Of three adjacent rows of carrots, the two nearest the Chives were large and perfect and the third row was better than average.

CHROMATOGRAMS   *(See* APPENDIX*)*

CINQUEFOIL *(Potentilla fruticosa L.)* Observations recorded by Krasil'nikov tell of the inhibitory action of the butternut, or gray walnut tree *(Juglans cinerea)* on the growth of the creeping cinquefoil. The latter is suppressed in the vicinity of this tree for a distance approximately twice the diameter of the foliage. The same applies to the black walnut.

The principle applies in the case of many trees, in many regions. How much time and effort and expense have been expended in New England trying to make the plants grow too close to a sugar maple tree!
(See COMPANION PLANTS, INTERCROPPING*)*

CITRUS *(Citrus)* Citrus trees like the protective influence of rubber trees, live oak and guava.

CLOVER *(Leguminosae)* Clover stimulates the growth of Deadly Nightshade *(Solanum nigrum)*. Clover is also a good plant companion to grow with all of the grasses for lawn or for pasture because the clover adds to the soil the nitrogen which is needed by the grass family. Clover is harmed by the presence of Henbane and members of the buttercup family. *(See* BUTTERCUP FAMILY*)*

COCCIDIOSIS Even in their use as food and medicine, certain plant combinations work better together than separately. Greens of Dandelions and young Stinging Nettle, given as feed help to prevent or cure coccidiosis in baby chicks. *(See* DANDELION, STINGING NETTLE*)*

COFFEE *(Coffea)* Strong coffee is good for digestive troubles of many animals, mostly the ruminants.

COMBINATIONS TO AVOID Some plant combinations are bad. Among the herbs, Rue and Sweet Basil do not thrive near each other. Fennel seems to be of the most doubtful character; it hinders germination of Caraway and Coriander, disturbs the growth of bush beans and blights the growth of tomatoes. Tomatoes and kohlrabi do not like to grow together. In fact most of the cabbage family should not be grown near tomatoes. The tomato has such an antipathy for quack grass *(Agropyron repens)* that it can be used to suppress it.

COMPANION PLANTS There are various reasons why certain plant combinations are successful. Plants having complementary physical demands are well suited to one another. For instance, a plant which needs plenty of light may be a good companion to another plant which can stand partial shade. Plants needing plenty of moisture may get along well with others which need less moisture. Deep-rooting plants open the ground for other species with less deep roots. Deep roots utilize a different part of the soil from shallow roots. Similarly, tall plants use a different part of the area above the garden from that filled with low-growing plants. Heavy feeders should be followed by light feeders, or plants that make the soil rich again, such as legumes. Plants that cannot stand the competition of weeds, should follow those that leave the soil free of weeds.

Gardeners and biochemists are still investigating other less obvious factors in its environment which may influence plant health, such as aroma, leaf and root exudations, or influences from the roots of still other plants further out in the environment.

(See INTERCROPPING, MAPLE, CINQUEFOIL*)*

COMPOST One of the principles of making biodynamic compost is that plants, parts of plants, extracts and decoctions can influence the fermentation process that should go on in the compost heap. Plant preparations can be used on purpose to organize the complicated fermentation process as it occurs in the process of making compost. Even when added in tiny amounts to big masses of rotting materials, such plant preparations can influence the whole pile. Once the conditions of moisture, aeration, and temperature are satisfactory, a compost pile will soon be inhabited by myriads of earthworms, enchytraeids, and other small, mostly microscopic animals. These will carry the effects of the bacterial life, stimulated by the inoculants, all over the pile, thus promoting controlled fermentation.

There are a few simple rules to keep in mind when making biodynamic compost. Compost heaps should not be built under conifers because the turpentine substances retard proper fermentation. Do not build a compost heap on top of grass sod; dig out the sods and add them to the compost heap, but let the heap itself be built on the soil directly. Grass sods will impede fermentation. *(For details an composting techniques, see the biodynamic literature.)*

Professor A. Seiffert, who is a long time promoter of compost making on the old continent, says in a past article *(March, 1964),* that the healing potentials in abundantly growing weeds can best be harnessed by putting them into the compost heap. This helps to combat what he calls "the exuberantly acting forces of nature." He reports that when he had to start a large vegetable garden some years ago on a heavy loam, he found the land completely overgrown with quack grass and Canada thistle. All the indicators of poorly drained, heavy soil, such as mint, buttercup, forget-me-not, knotweed and dock were also there. Large quantities of thistle and quack grass *(Agropyron)* were used to make compost. When this was turned back to the land, thistle and quack grass lost their vigor. The soil is now looser and the weeds can be controlled with ordinary means. He even says that by regular return of weed compost back to the land he was able to raise a sensitive variety of early potatoes on the same piece of land seventeen times. The weed composts had steadily rejuvenated the soil.

COMPOST PLANTS In biodynamic composting, a set of six herbs, each of which is specially prepared, is used to control and influence the fermentation of composts and manures. These herbs are: Stinging Nettle *(Urtica dioica),* already known from everyday observation to be an excellent compost maker and soil builder, Dandelion *(Taraxacum officinale),* the bark of the oak tree *(Quercus alba and Q. rubra),* Yarrow *(Achillea millefolium),* Camomile *(Matricaria chamomilla),* and Valerian *(Valeriana officinalis).*

A detailed study of these herbs and their preparation would go beyond this presentation. Information about the BD Preparations and their application, is given by the Biodynamic Farming and Gardening Association, 25844 Butler Rd., Junction City, OR 97488.

CONIFERS *(Coniferae)* Dense conifer plantings should be avoided in the vicinity of compost heaps as being inimical to the fermentation process.

CORIANDER *(Coriandrum sativum)* Coriander hinders the seed formation of fennel. Coriander seed, when sown with Anise seed helps the latter to germinate. It also improves the growth and seed formation of the Anise plants. Bees and other beneficial insects are attracted to the garden while coriander is in bloom.

CORN *(Zea mays)* Sweet corn does well with early potatoes. It is also aided by beans and peas, which help the soil by putting back nitrogen which the corn uses up. Beans benefit from the slight shade given by the corn plants.

Other plants which appreciate the sheltering shade of corn are melons, squash, pumpkins, and cucumbers. The presence of a border of cucumbers is in turn beneficial to the corn. Rows of corn in windy areas help to break the wind until a good hedgerow has grown up.

We are told that when the early settlers came to this country, they found the Indians growing corn and pumpkins together; perhaps one of the earliest native American examples of Companion Planting.

CORNFLOWER *(Centaurea cyanus L.)* Cornflower in small quantities is beneficial in fields of small grains. Other common names for this annual are Bluebottle and Bachelor's Button.

COWS Marjoram is fed to cows to prevent abortion. Marjoram and balm tea is given to cows after calving. Coffee is a digestive aid. For many years European farmers have been using Biodynamic Cow Tonic as a dairy feed supplement to aid digestion, and improve the general health of the animals. The recipe for this herbal mixture came from a European farmer. It contains 23 herbs, the main ones being Stinging Nettle, blackberry and raspberry leaves, linden flowers, Caraway, Fennel, birch leaves, and Angelica root. It is used either in the form of a tea sprayed onto the feed just before eating, or as a dry mixture added to the feed, at the rate of one teaspoonful per cow per day. More information about this Cow Tonic may be obtained from the Secretary of the Biodynamic Farming and Gardening Association, Inc.   *(See* COFFEE*)*

CRAB GRASS *(Digitaria sanguinalis)* This troublesome grass is especially pestiferous when it gets into the lawn or garden. In the lawn it may be controlled by hand pulling before it gets a chance to form a mat. Good fertilizing, and not cutting the grass too short in the dry part of summer will benefit the good lawn grasses, crab grass cannot stand their competition. In the garden it can be controlled by frequent cultivation.

CROP ROTATION Biodynamic gardeners follow a system of crop rotation which can be described briefly as follows: The heavy feeders are planted immediately after fertilizing with a manure-containing compost. These include all of the cabbage varieties, cauliflower in particular, all leaf vegetables such as chard, head lettuce, endive and spinach, as well as celery and celeriac, leeks, cucumbers, squash and sweet corn. Rhubarb and tomatoes are also heavy feeders, but are not included in the crop rotation since rhubarb is a perennial, and tomatoes prefer to be retained in the same place year after year, and fertilized with tomato compost. There is a limit, however, as after seven to eight years, one may find the tomatoes showing signs of blight. The best remedy is to move them somewhere else.

The heavy feeders are followed by legumes or the soil improvers which add nitrogen to the soil through their root nodules. To this class belong beans and peas of all varieties.

The legumes should be followed by the light feeders which should be fertilized with a good compost, well-decomposed. They include the bulbs and all root vegetables such as carrots, beets, radishes, salsify, parsnips, turnips and rutabagas. For details, check on cultural practices in garden books.

CRYSTALLIZATION In addition to observations of nature and planting experiments in garden and field, there exists still another way of finding out about various plants likes and dislikes of their plant neighbors. This is through the application of the *Sensitive Crystallization Method*. This method, originated by Dr. E. E. Pfeiffer (following up an indication given by Dr. Rudolf Steiner), is already widely applied to testing finer, dynamic forces and qualities in the plant, animal, and human realms.

Dr. Pfeiffer invented his Sensitive Crystallization method in the course of a line of research suggested to him by Dr. Rudolf Steiner. The latter had asked him to find a reagent, that would reveal the "formative forces" in living matter. He first used Glauber's salt *(sodium sulphate)*, for crystallizations. He then used copper chloride, with additions of extracts of living matter, and got excellent results. He also investigated a great number of other chemicals in this connection but found that copper chloride was by far the most suitable medium. Thereupon he began developing what has now become known to the world as The Pfeiffer Method of Sensitive Crystallization.

It is being used in a number of laboratories in various parts of the world, and is important for quality tests of food products, organic fertilizers and fertilizer aids, seed crops to be used for planting, and in fact any field where *quality* is essential. The same crystallizations have been worked out as blood tests, to aid in early medical diagnosis.

The method consists of adding small measured amounts of the substance to be tested – for instance, a plant extract – to 10ml of a 5% solution of copper chloride in a test tube. After thorough mixing, this solution is poured out onto perfectly smooth, specially cleaned, round glass plates of about 9CM diameter, which are provided with a rim. These plates with the solution are then left in a temperature and moisture controlled room on a leveled surface. The solution slowly evaporates, and after 14-17 hours, it crystallizes into a pattern, which is determined by the nature and quality of the plant from which the extract was taken. Those forces (or factors), which are inherent in the plant and bring about its specific form and shape *(formative forces)*, as well as the living growth forces, are active in forming the resulting pattern of crystal arrangement.

A strong, healthy, vigorous plant will produce a beautiful, harmonious, and clearly formed crystal arrangement, radiating through to the outer edge.

On the other hand, a crystallization made of an extract from a weak, sickly plant will result in a picture which shows thickening or incrustation, be more or less uneven, unharmonious. These results were confirmed with thousands of crystallizations made in the years from 1927 to 1938. In consequence, one can now say, whenever one sees such a clear, harmonious plant pattern, the plant used must have been healthy and of good quality.

This crystallization test has proved to be adapted to practical application in various fields to determine the inherent quality of a living organism. For example, it can be used as a test of the influence of different treatments in preserving food plants. Among its many applications has been its use as a test for the relationships between plant and plant. "Sensitive" as it is, it can serve as a reagent for the exceedingly fine and subtle forces active in a plant symbiosis. The procedure is to make crystallization pictures of each of the two plants to be tested as to their sympathetic or antipathetic relationship. Then a third crystallization is made with a mixture of the extracts from these two plants. This third picture is the determining one.

The determinant crystallization can present several possible aspects for interpretation. Let us take cucumbers and beans as a concrete example. We have the original pictures of the bean and the cucumber, each showing its specific and characteristic pattern. Now the crystallization of the mixture of both may result in any of the following: **A)** An undisturbed, and even stronger cucumber picture, which would mean that the bean enlivens and strengthens the cucumber in its growth. **B)** The mixture may show a stronger bean picture, then the cucumber would be the helping, strengthening plant. **C)** The picture may show a balance of both plant signatures in a harmonious and clearly formed pattern; this would indicate a mutual and strong beneficial relationship between these two plants. **D)** If the resulting picture shows deterioration in its form pattern, in comparison with the two original ones, and is thickened or encrusted, unharmonious, giving a disturbed impression, then this indicates a mutual antagonism.

In testing vegetables we use mostly the seeds, but also roots, leaves, and sometimes the whole plant. For the study of the interrelationships in forestry, the extracts from the roots of different trees and shrubs gave the most revealing results in the crystallization pictures. However, extracts of seeds and borings from the trunks were also used, and gave further material for evaluation.

One gram of the plant material (*e.g.* peas) was finely chopped and 20cc distilled water added for softening. This was then gently mortared, and let stand four hours for extraction; then filtered through cloth. Of this extraction, 1cc was added to 9cc of a 5% solution of copper chloride. We also used a mixture of 4cc of the extract, to 4cc of a 10% solution of the copper chloride. This latter concentration results in pictures more typical for the plant species as such, while the 1-to-9 concentration reveals the activity of the finer forces active in the plant. For the crystallization of the mixture of the two plants to be tested for their specific relationship, two different combinations of the plant extracts were used: **A)** equal parts of both, **B)** one plant extract in one tenth of the amount of the second plant extract. The first gives insight into the favorable or unfavorable relationship between the two plants in general. The second mixture may result in a good picture even though an unfavorable crystallization resulted from the mixture of equal parts. This is important and indicates that a plant should be used as a border plant only, and as such can have an especially good and far-reaching influence.

The following few examples have been selected from the countless series of crystallizations made, to illustrate how the results of such tests can confirm observations and field tests. Many of the series paralleled plot and field experiments with the same plants. The plants were observed while growing, and were later measured and weighed, with the results compared to the crystallization pictures.

In the case of the already mentioned *bean and cucumber* relationship, the crystallization of the mixture of equal parts of both plants showed a predominant but harmonious and stronger cucumber pattern, in comparison with the cucumber picture alone. This indicates that the beans improve and strengthen the growth of the cucumber if they are planted next to them, or preferably as a border.

Peas and cucumber, showed a similar result to that described above, but not quite so pronounced. The mixed extracts of *peas and carrots,* in equal parts, gave a harmonious interpenetration of the two original pictures with their clear and finely branched crystal patterns, revealing thus mutual benefit.

Beans and fodder turnips: The beans alone gave a finely branched crystal pattern, radiating through to the border. The picture of the turnip, on the other hand, had rather dense and stiff crystal arrangements, more contracted towards the center of the plate. The mixture now of both resulted in a harmonious picture with the form characteristics of the turnip predominant, but loosened up, enlivened into a clearer and more refined pattern. The beans benefited the turnips.

Kohlrabi and tomato: Both plants crystallized alone gave a beautiful and clearly defined plant pattern. But the mixture of the two resulted in a rather disorderly and partly smudged pattern, contracted, thickened and not radiating through to the border of the plate. From this we can learn that these two plants dislike growing near one another.

Wheat and field poppy: The mixture of these two plants used in equal parts as well as in the proportion of 1 poppy to 10 wheat, gave in both cases unfavorable pictures. The clear, beautiful, and sharply defined plant pattern of the wheat appears in the mixture, broken up into small, dense clusters of crystal groups, more or less isolated from each other, a formation which we learned to know as a sign of disorganization, deterioration.

Wheat and esparsette: In this association the mixture of equal parts of both plants showed a slightly changed but still clear and typical wheat picture. A beautiful and even more refined wheat picture resulted from the mixture of 1 esparsette *(sainfoin),* to 10 wheat, thus indicating that esparsette as a border plant around the field has a very beneficial influence. *(See* BORDER PLANTS*)*

There is another method of testing plant relationships which has been used and is now being further developed by Dr. Koepf. This consists of growing a plant directly in more or less diluted extracts of another plant in order to determine the specific influence of this second plant on the first. These experiments are carried out in the laboratory under controlled and equal conditions of light, temperature, etc. With this method the finest variations in concentration, different parts of the plants, etc. can be used in the tests, and the resulting growth observed, and finally measured and weighed. This promises to give ever more and better insight into these most interesting and so-important relationships between plant and plant.

[Erica Sabarth]

CUCUMBER *(Cucumis Sativus)* Cucumbers appreciate some shade. They grow well in the corn field in alternate rows. They also grow well in alternate rows with early potatoes and early cabbage. A few radish seeds sown in the hill repel Cucumber Beetles. Borders around the cucumber patch can carry companion plantings of kohlrabi and lettuce, kohlrabi and early savoy cabbage, celeriac and lettuce or bush beans, or lettuce and radishes. A border of beans helps the cucumbers, and the latter in turn, help the beans. Cucumbers do well with sunflowers.

Cucumbers are especially helped by compost made of horse manure and sods, best of all, containing couch grass.

If the cucumber vines are attacked by downy mildew, it will appear as yellowish-brownish spots with purplish-grey mildew on the lower side. The disease may be prevented by using Stinging Nettle liquid.
(For directions, see BIODYNAMICS #64, *pp.29-36.)*

Krasil'nikov cites experiments, showing that potatoes grown close to cucumbers were more likely to be affected by Phytophthora Blight. *(See* CHIVES*)*

CUTWORMS *(Noctuidae)* *(See* MULCH*)*

CYPRESS SPURGE *(Euphorbia cyparissias)* Cypress Spurge is known to have a very bad effect on grapes, causing the vines to be sterile. This is a useful warning to everyone who has grapevines in areas where Spurge is a very common weed or has been grown as an ornamental. Cattle eating hay containing this plant may be seriously poisoned.

DANDELION *(Taraxacum officinale)* Dandelions exhale ethylene gas which inhibits the height or growth of neighboring plants. They also cause flowers and fruits of neighboring plants to mature early. Dandelions have a special affinity for alfalfa. *(See* APPENDIX*)*

DATURA *(Datura stramonium)* Commonly called Jimson weed; also called Jamestown Lily, Dewtry, Thorn Apple, Devil's Trumpet or even Barnyard Jasmine. Has been observed by nurserymen of recent years to protect nearby plants against Japanese Beetles. All parts of this plant are poisonous to livestock and humans.

DEAD NETTLE *(Lamium amplexicaule, L.)* Usually called Henbit; sometimes called Blind Nettle or Bee Nettle, has no relationship to Stinging Nettle. It is a mild looking, unassuming member of the mint family, about six to eight inches tall with a white flower that might look like a tiny weasel if one views it with imagination, hence the name Lamium which means weasel. A few plants in a border will aid all vegetables.

DEADLY (BLACK) NIGHTSHADE *(Solanum nigrum L.)* Black Nightshade, as this plant is often called, or Poison Berry, does not grow too well in the presence of Wormwood or White Mustard but its content of poisonous alkaloid in the berries is correspondingly enhanced. Clover, on the other hand, stimulates the growth of Nightshade. This plant should not be confused with Belladonna *(Atropa)* often called by the same name, Deadly Nightshade.

DILL *(Anethum graveolens)* Dill may be sown in very small quantities in corners of the vegetable garden where it may be allowed to mature and bloom for the honey bees. It is also a good herb to grow with cabbages. Light sowings are often made in carrot, lettuce, onion or cucumber beds. **WARNING**: Dill sown with carrots, *greatly reduces* the carrot crop if the Dill is allowed to mature. Therefore, it should be pulled out *before* going to bloom. The suppressing effect of Dill on carrots and tomatoes is evident, even if sown in the ratio of 20-to-1.

EGGPLANT *(solanum melongena)* To protect eggplant from the Colorado potato beetle, plant it among green bean plants. The potato beetles prefer eggplant, even to potatoes, but the green beans repel the potato beetles in the eggplant rows, as much as in the potato patch. A border of eggplants around the potato patch will concentrate potato bugs, where they may be caught and controlled.

ELDERBERRY *(Sambucus nigra* AND *Sambucus canadensis)* Elderberry shrubs growing around the sites of compost and manure heaps assist in fermentation of the compost. Humus under Elderberry bushes has long been known to be especially light and fluffy, and such humus can be of great value added to topsoil in the garden.

ELM *(Ulmus)* Grapevines climbing on elm trees bear especially good grapes.

ESPARSETTE *(Onobrychis viciaefolia)* This is a Eurasian, pink-flowered, perennial forage legume, also known as Sanfoin. Esparsette aids most vegetables, if planted only in the border of the vegetable plot. It is a plant which thrives on calcareous soils. Dr. Steiner recommends it also as a border plant for small grains, or as a lightly scattered stand mixed with the grain. The seed keeps its germinating power about two to three years. Three pounds of Esparsette seed per acre, mixed into seedings of small grains or corn, will provide a scattered stand of this plant.

EUPHORBIA *(Euphorbiaceae)* In general, the Euphorbias – or spurges – help and protect tender plants of other species, because they foster a soil which preserves warmth, thus simulating conditions of a warmer climate. Euphorbia is a vast family in other parts of the world; with over 4000 species.

Euphorbia lactea, otherwise called the Mole Plant, is said to repel moles, mice and rats. If it is planted near young trees it may keep mice from girdling the trunks. Sow the seeds in late fall to germinate in early spring. Sown at other times, they do not germinate well nor get to sufficient size to repel the mice. Sow two to three seeds, about 40' apart, around the garden. Be sure to mark the seeds with stakes, so as not to destroy the seedlings. These are hardy-biennial.

Another member of this family, called Spurge *(Euphorbia lathyrus)*, also repels mice and moles.

Cypress Spurge *(E. cyparissias)*, is inimical to grapes.

(See CYPRESS SPURGE*)*

FENNEL *(Foeniculum vulgare)* Fennel is the exception to the rule, that most herbs have a good influence on plants in their vicinity. Fennel has a harmful effect on bush beans, Caraway, tomatoes, and kohlrabi.

Contrariwise, Fennel seed formation is prevented by the presence of Coriander. Furthermore, Fennel suffers still more, in the presence of wormwood. A Fennel plant three inches from Wormwood was only one-seventh the height of Plants four feet away. Finely chopped leaves of Wormwood in a seedbed reduced the germination of Fennel to one half. Analysis of raindrops on the leaves of Wormwood showed that they had absorbed a high proportion of absinthin, which affected the germination of the Fennel seed. *(See* HERBS*)*

FERNS *(Pteridophyta)* Ferns and beeches *(Fagus)*, do not like each other. Englishman, M. C. Rayner, found out that compost made from fern is favorable to tree seeding. It can successfully be used in tree nurseries, encouraging germination.

FLAX *(Linum usitatissimum)* Flax acts as a smoothing harrow on crusty and lumpy clay soils, but should never be planted in the same place more than once in five years. It is a very useful crop to prepare the soil one year, for raising carrots the next year. It is also possible to improve a carrot crop by alternating two rows of carrots with one row of flax. Decomposition of organic matter in the soil is favorably influenced by flax, and the soil structure is left in excellent condition especially in the upper layer.

One or two plants of flax to a row of potatoes will remarkably reduce the number of Colorado potato beetles.

Flax and linseed are adversely affected by leaf-extracts of the Camelina species, which is a harmless appearing weed often found in flax fields, in Europe and America. In fact this weed is so common in these fields, that it is known as False Flax *(Camelina sativa* AND *C. microcarpa)*
(See CARROTS, POTATOES*)*

FLEA BEETLE, BLACK *(Haltica and Epitrix)* Flea beetles on radish are checked by inter-planted head lettuce. Flea beetles in radish and kohlrabi have been controlled by sowing the rows in the following sequence: lettuce – radish – kohlrabi – radish – lettuce. Flea beetles are also repelled by Wormwood and Mint. These creatures like a dry, crusted soil. Light cultivation and increase of humus content of the soil help to discourage them.

FOXGLOVE *(Digitalis purpurea L.)* Foxglove has a growth stimulating effect on other plants, and gives an enduring quality to plants growing in its neighborhood. It likes the borders of the forests, and open places in the woods. It seems to influence the growth and duration of stands of trees. Foxglove and pine trees are good companions. To make cut flowers last longer in a vase, try adding some tea made of foxglove. *(See* PRESERVATIVE EFFECT*)*

FRENCH MARIGOLD *(Tagetes patula)* In the public parks of a city in Holland, instead of chemical sprays or soil treatments to take care of tiny, nematode worms infesting the soils within rose gardens, the Dutch Plant Protection

Service implemented French Marigolds (actually an import from Mexico, despite the name). The marigold excretes a substance from its roots, which kills soil nematodes. Marigolds planted among the rose beds controlled the nematodes, while rose beds without still suffered from infestation. This story is told by Rachel Carson in Silent Spring, and she also adds that in many places gardeners now know how to plant marigolds to combat nematodes in other crops.

A friend in Ohio writes of visiting a garden, where gardeners have long made it a practice to grow marigolds between tomato plants. They report that tomatoes grow better, and bear more fruit with marigolds than without them. Marigold is successfully used to combat potato nematodes. French Marigold is a cure for white fly in tomatoes, and is used in greenhouses. The odor of marigold foliage, and blossoms of the old type, is effective as an insect repellent. Marigolds have been successfully used in bean rows to repel the Mexican bean beetle.

FRUIT TREES All fruit trees do well when treated with a threefold paste of cow manure, very fine diatomaceous earth, and clay, mixed with Equisetum tea.   *(See* HORSETAIL*)*

The mixture should be applied to the stem and thicker parts of the branches with a whitewash brush or sprayer, before winter, and early in spring. The bark should be brushed and cleaned before the mixture is applied. In addition, Biodynamic Preparation 500 is sprayed on the ground, and B.D. 501 on the trees before the blossoms come, or when the fruit is set.

Fruit trees are aided by intercropping of 15% mustard with legumes. Other beneficial companion plants for fruit trees are Stinging Nettle, Garlic, Chives, Tansy, Horseradish, Southernwood, and Nasturtiums.

GARLIC *(Allium sativum)* Garlic promotes the growth of vetch. Garlic and roses have a mutually beneficial effect. Garlic, onions and shallots inhibit the growth of peas and beans. European peasants used to put pieces of garlic into their grain for protection against weevils.

Tea made of garlic, onion or chives may be used to control such severe diseases as late blight on potatoes and tomatoes. It is also used against brown rot of stone fruits. Use the tea shortly after its preparation.

(See CHIVES, ONIONS, FRUIT TREES*)*

GERANIUM *(Pelargonium)* White geranium is said to be a catch plant for Japanese beetles. They are attracted to it, eat it, and are then said to die.   *(Research suggested)*

GERMINATION Seed germination of many plants is inhibited by substances called cholines and blastocholines, *(Gr. Blastos = sprout* AND *cholycin = to prevent),* inhibiting germination of their own seeds, and also seeds of other species of plants. They may be excreted by the roots of seedlings, and suppress growth of neighboring plants. Root excretions of seedlings of wheat and rye grass *(Lolium perenne),* suppress germination of seeds of field or corn Camomile, scentless Camomile, or corn Mayweed *(Anthemis).* Seedlings of beans suppress germination of seeds of flax and wheat, while seedlings of violets inhibit germination of wheat seedlings.

Summing up research carried out from 1932-1955, Krasil'nikov says that various volatile substances are present in many odorous plants. In some plants they are formed in the seeds and fruits, while in others in the leaves and stems or in the roots. Essential oils of a series of plants: citrus plants, Clove, Mint, Summer and Winter Savory, Thyme, Germander, Eucalyptus, etc. and the resin of coniferous trees, poplar and others inhibit the germination of seeds of various plants to various degrees. *(See* MORNING GLORY*)*

GLADIOLUS *(Gladiolus)* The Greggs reported a very skillful market gardener, saying that gladiolus is strongly adverse in its effect upon peas and beans, and the influence can be noted as far away as fifty feet or more.

GOOSEBERRY *(Grossularia reclinata)* In 1950, a study was made of the lethal effect of the volatile substances from tomatoes on certain insects attacking the gooseberry shrub. The observer, quoted by Krasil'nikov, noted that gooseberry shrubs in the vicinity of tomatoes were protected from certain insects. *(See* CHIVES*)*

GRAPEVINE *(Vitis* AND *Muscadinia)* A grapevine supported by either an elm tree, or a mulberry, will grow well. The yield of grapes is increased by nearby Hyssop. Grapevines are also aided by 15% mustard, with legumes as intercrop.

Cypress Spurge is inimical to grapes.

(See CYPRESS SPURGE*)*

GRASS Grass growing under fruit trees somewhat suppresses root growth of apples and pears. Some grasses have a negative influence on one another. It was found in experiments with mixed pastures that when timothy was grown with spear grass, meadow grass with spear grass, and meadow grass with timothy grass, the yields were much lower than when these grasses were grown separately.

Grass growing as a weed sometimes has an adverse effect on neighboring plants. An article in the ANNUAL OF APPLIED BIOLOGY, 1947, *The Competition Between Barley and Certain Weeds under Controlled Conditions,* states that velvet grass *(Holcus mollis)*, slows down the growth of barley. *(See* APPLE TREES*)*

GROWTH REGULATORS Part of the science of plant study now includes investigations of substances which can be harnessed to accomplish certain, definite results in our plant world. Many of these compounds are isolated from plants, fungi, peat moss etc. Some are synthetics. Some of the substances are called *auxins*, from the Greek word, which means *to increase*. These auxins are used in horticulture and in agriculture for the following purposes: to control the plant's habits of reaching towards the light, sending roots down into the earth, developing top growth and forming fruit. In practical use, growth regulators are applied to insure and speed up root formation on cuttings, to prevent potatoes from sprouting, to prevent leaf drop, and to defoliate certain plants before harvesting. They can also be used to inhibit internode elongation − to reduce the need for trimming ornamental plants and lawns, to force fruit set without pollination − as in seedless tomatoes and to regulate the flowering of pineapples, so that the crop harvest can be staggered, and the size of the fruit increased.

Gibberellic compounds have been used to stimulate internode elongation and gibberellic sprays on grapes change the size and shape of the fruit. It must be emphasized, that in spite of sometimes striking effects, these substances cannot outweigh the significance of the basic growth conditions − which are sunlight, temperature, air, water, and nutrients. The specific effects of companion plants are partially due to the presence of biotic substances in a plant. The effects exerted this way from one species to the other are relatively mild.

Applications of isolated and concentrated effective compounds or synthetic compounds may strikingly upset the harmonious growth of the plant in space and time, in other words disturb their shape and growth rhythm.

GUAVA *(Psidium guajava)* The guava tree protects nearby citrus trees.

HAZELNUT *(Corylus)* It is good to have hazelnut trees and bushes in the fence rows of pastures. The cattle like to nibble on them. Leaves of hazelnut when eaten by cows increase the butterfat content of their milk. The tannic acid in the leaves also has a cleansing effect on their digestive system. Cows are also fond of resting under nut trees of all kinds; they are less troubled with flies there. Hazel shrubs are good in mixed hedges around compost yards.

HEAVY FEEDERS *(See* CROP ROTATION*)*

HEAVY SOIL Rape and soybean loosen heavy soil and leave it friable. Flax favorably influences decomposition in the soil and puts the soil structure in excellent friable condition, especially in the upper layer. Buckwheat loosens heavy soil. Caraway loosens heavy soil and puts the ground in good physical condition.

HEDGE MUSTARD *(Sisymbrium altissimum)* Sometimes called Blister Cress and Tumble Mustard. Likes to grow near oats. This mustard has an alkaline secretion around its roots which sweetens the soil. Inimical to turnips; in Europe they say, "Hedge Mustard eats the turnip."

(See MUSTARD*)*

HEDGES Provide a windbreak which is significant in areas where dry wind blows away soil moisture vapors and

carbon dioxide exhalations from the ground. Useful birds and other animals find a place to live. Bees like flowering hedges. *(See* BEES*)*
See also special literature on how to plant and maintain a hedge.

HEMP *(Cannabis sativa)* It has long been known from European gardens that a border of Soft Hemp – as it is generally called – around a cabbage field will keep away the white cabbage butterfly. Now it has been discovered that this hemp has a protective effect on plants growing in its vicinity because of volatile substances which it excretes. These volatile substances inhibit the growth of certain pathogenic microorganisms. Unfortunately it is now practically impossible to procure Soft Hemp seed any more. Another name for this plant is Marijuana, which may explain why. The other fiber hemps, Abaca, Sisal and jute, are not related.

HENBANE *(Hyoscyamus niger, L.)* Henbane is harmful to clover. Poisonous to fowls, hence the common name. Hogs are killed by eating its fleshy roots, the ripe seeds kill chickens. All parts of the plant contain poisonous alkaloids.

HERBS Since time immemorial, the various aromatic herbs have been planted as a border, or in small patches in vegetable gardens, where they are known to be beneficial to the more stolid vegetable plants. The one exception to this general rule is Fennel, which has an adverse effect on several plants. *(See* FENNEL*)*

Many herbs have a good influence on plants in their vicinity. All vegetables are aided by most aromatic herbs, e.g. Borage (best grown in a nearby corner or in a separate bed), Lavender, Hyssop, Sage, Parsley, Chervil,

Tarragon, Chives, Thyme, Marjoram, Dill, Camomile, Lovage; but *not* Wormwood or Fennel. If these are planted around the borders or at the ends of raised beds, they will sometimes help repel certain insects like the cabbage butterfly. Other herbs, including Santolina, Winter Savory, Blessed Thistle, Blue Hyssop (sometimes Pink and White), Lavender, and Marjoram are said to repel certain insects. Experiments are needed to make this information more exact, and records should be kept. Onions, shallots and garlic also repel insects.

Another beneficial effect of herbs is that they bring an aromatic scent into the atmosphere when planted amongst vegetables, in small proportion, as in a border or at the end of the row.

Stinging Nettle growing near any herb plant will increase the pungency and aroma of the herb itself. Stinging Nettle growing near Peppermint nearly doubles the quantity of essential oil in the Peppermint.

(See PEPPERMINT*)*

Yarrow also increases the aromatic quality of all herbs which grow nearby. *(See* YARROW*)*

Mixed cultures of herbs make faster and closer growth than a single species. On very poor stony soil the following species, when intercropped, formed a very good, close stand within a short time: Rue *(Ruta graveolens)*, Hyssop *(Hyssopus officinalis)*, Sage *(Salvia officinalis)*, Lavender *(Lavendula spica)*, Thyme *(Thymus vulgaris)*, St. Johnswort *(Hypericum perforatum: H. vulgare)*, and Southernwood *(Artemisia abrotanum)*.

Another way in which herbs help build and maintain good gardens is by controlling biologically, both insect pests and plant diseases. Here again prevention is better than cure. Not only individual plants but whole landscapes become diseased through monocultural practices, since nature, left to herself, never produces acre after acre of only one kind of plant. Usually the more variation the better, whether in general landscape development, a farm unit, or a garden. In medieval horticulture no lines were drawn between the flower, vegetable and herb gardens. All grew together to their mutual benefit. Now, in the plant residues, root excretions and leaf and flower emanations absorbed from one plant to another we can see a reason for these benefactions. They are connected with the delicate balances which normally exist in countless numbers in nature, and which man unwittingly disturbs.

The influence of one herb upon another is another aspect of conservation and good gardening which relates to the quality of produce and the health of consumers.

Teas are a good means to transfer neighborly impulses from one plant to another, as Dr. Pfeiffer explained it. When the leaves carry the active principle, the plant for tea must be gathered early in the morning, not later than ten o'clock and before blooming – while the blooming process is still found in the leaves – before it has dissipated into the blossom. It has been proved in this case that older plants do not have the same strong effect. Dry in the shade and store in a dry place, preferably in tin boxes. For many herbs like Camomile, Yarrow, St. Johnswort, and others, the blossoms are the part used.

To make an herb tea to be used as a spray, cover the medicinal plant with water in a pot, bring just to the boiling point, and take off the fire. This infusion should be diluted with four parts water. It is recommended that the fluid be stirred for ten minutes. It should then be used immediately.

Stinging Nettle tea will combat plant lice or aphis.

Chives tea is useful to overcome apple scab. Use dried Chives. Do not boil Chives but instead pour boiling water over dried Chives and leave it to infuse for 15 minutes. Dilute the infusion with two or three times as much water, and stir. The best results have been obtained with a comparatively strong solution.

Chives tea is useful to combat gooseberry mildew.

Horse-radish tea fights monilia in fruit trees. Use young horse-radish leaves at the beginning of the attack. Make the tea as explained above, and dilute with four times as much water.

Most vegetable plants have inconspicuous flowers, or in the case of root crops, no blossoms at all the first year. This makes for a one-sided atmosphere among the vegetable plants, lacking in the realm of blossoms formed by warmth and light. If the garden is surrounded by borders of summer flowering plants of mixed varieties, the latter will attract a wide and balanced variety of insects, thus promoting pollination of all neighboring vegetation. Some of the summer flowering plants and shrubs which have a beneficial effect are Wild Rose, Elderberry, Buddleia, Privet, Goldenrod, and Bee Balm.

Some herbs may be scattered about the garden, for instance, one herb clump at the end of each raised bed, to help overcome the monoculture, and to create a lively aromatic atmosphere in the bed where it is growing. Small herb hedges may be grown, for instance, of Hyssop, or of Lemon Balm. When Hyssop is in bloom, it is bedecked with all manner of moths and butterflies, wild honey bees, and other flying insects, all of which benefit the whole garden. Parsley and Dill, Coriander, and Bee Balm, allowed to blossom, will provide a welcome for honey bees and butterflies, who bring their good influence to the heavier vegetables which are confined there.

HONEY BEES *(Apis meffifera)* Are especially attracted to Thyme, Catnip, Lemon Balm, Pot Marjoram, Hyssop, Sweet Basil, Summer Savory, and Mint.

Whenever bees swarm, we take a generous handful of Lemon Balm and rub it into the inside of the new hive to make the new swarm remain there.

HORSE-RADISH *(Armoracia rusticana)* Aids potatoes if it is restricted to only the corners of the potato plot. There is a symbiotic effect between the two plants while they are growing, and the potatoes are more healthy and resistant to disease. This does not necessarily mean that the horse-radish inhibits potato bugs.

Dig up horse-radish plants after each season, or it will spread too far. Try to remove all of the roots and pieces of root or they will start growing again and will spread. The condiment for the table is made by grating these dug up roots and adding vinegar – a very tearful process unless it is done outdoors in a fair breeze.

Horse-radish tea is effective against monilia on apple trees.
(*See* HERBS, APPLE TREES)

HORSETAIL *(Equisetum arvense L.)* The dried leaves and stems of the sterile form of this plant, often called meadow pine because of its resemblance to pine trees, are made into a decoction and sprayed on plants to be protected against different kinds of fungi. *Equisetum arvense* sends up fertile spore-bearing stems from April to May. These die down shortly and the stems used for the tea spring up from the same perennial, creeping, branching rootstocks. These bear a high percentage of silica which has a controlling effect on fungus diseases.

To make Equisetum Tea, place 1 1/2oz. of dried Equisetum in one gallon of cold water. Bring it to a boil and let it boil for 20 minutes, but no longer. Allow to cool gradually. Strain. Spray on all plants which have their true leaves developed. Dilute the decoction more, each time it is used in the same place.

(See BIODYNAMICS, *#54, p.26, for further suggestions as to the use of similar plant sprays)*
(See SILICA*)*

HYSSOP *(Hyssopus officinalis)* Hyssop planted near grapevines increases the yield of grapes.

Hyssop tea may be used on the plant diseases caused by bacteria.

Hyssop attracts the cabbage butterfly and lures it away from the cabbages.

Radishes are not happy if Hyssop is too close by.

Both Pink and White Hyssop, but mostly the blue, are said to be insect repellent.

(See HONEY BEES, GRAPE VINES*)*

INSECT PESTS Insect pests, when they get out of hand, are one sign of upset balances in nature. Restore the right proportions in nature and pests gradually become less troublesome. This does not mean we shall ever be without insects. In a balanced scheme they are a necessity but because of man's interference their destructive powers occasionally become abnormal. Even more disturbing today is man's economic insistence that agricultural products must be large and unblemished by insects. One of the tasks of the biodynamic farmer and gardener today is to restore conditions which approximate those of nature without using unnatural substances. Making use of plant combinations which repel troublesome insects or which attract helpful insects is one part in this restoration of natural conditions.

INSECT PESTS AND PLANT CONTROLS *For specific plants to control or repel specific insects, look under plant named.*

INSECT	PLANT ANTAGONIST
ANTS	Spearmint, Tansy, Pennyroyal
APHIS	Nasturtium, Spearmint, Stinging Nettle, Southernwood, Garlic
BEAN BEETLE, MEXICAN	Potatoes
BLACK FLY	Intercropping, Stinging Nettle
CABBAGE WORM BUTTERFLY	Sage, Rosemary, Hyssop, Thyme, Mint, Wormwood, Southernwood
CUCUMBER BEETLE, STRIPED	Radish
CUTWORM	Oak leaf mulch, tanbark
FLEA BEETLE, BLACK	Wormwood. Mint
FLIES	Nut trees, Rue, Tansy, spray of Wormwood and/or tomato
GRUB, JUNE BUG	Oak leaf mulch, tanbark
JAPANESE BEETLE	White Geranium, Datura
LICE, PLANT	Castor bean, Sassafras, Pennyroyal
MOSQUITO	Legumes
MOSQUITO, MALARIA	Wormwood, Southernwood, Rosemary,
MOTHS	Sage, Santolina, Lavender, Mint Stinging Nettle, herbs *(See* APHIS*)*
POTATO BEETLE, COLORADO	Eggplant, Flax, Green beans
POTATO BUGS	Flax, Eggplant
SLUGS	Oak leaf mulch, tanbark
SQUASH BUGS	Nasturtium
WEEVILS	Garlic
WOOLLY APHIS	Nasturtium
WORMS IN GOATS	Carrots
WORMS IN HORSES	Tansy leaves, mulberry leaves

INTERCROPPING When several crops are grown on the same space, they are intercropped, and harvesting takes place at different times. After the first crop is harvested, the second crop remains as a ground cover until it is ready for harvest. With a crop like broad beans it is a great help to grow an intercrop, like spinach which shades the soil and prevents a cracked and crusted soil surface which is attractive to black flies. Spinach also helps maintain the soil microlife and moisture. Spinach is rich in saponin which is of benefit for following crop like cabbage.

Observation of external requirements, and our experience teaches us not to put a light-demanding plant near a tall growing plant with broad, shade-giving leaves, nor would we put a moisture-demanding plant with another equally demanding of moisture. Besides their particular nutrient requirements, plants are chemically affected in four ways that we now know about:

1. by aroma
2. by exudations from the leaves and roots
3. by the roots of other plants
4. by soil microorganisms

Here are some further examples of intercropping, according to F. Caspari: Bush or dwarf peas, when planted with early potatoes, grow poorly; the neighborhood of carrots enhances their growth. For a long time, Dutch gardeners have grown leek and onions with carrots. Onions are also friendly to red beets, strawberries, and tomatoes. In good soils, some Camomile scattered between onions is said to be helpful to onions. Growing bush beans adjacent to onions is not good practice. Celery does better when intercropped with members of the cabbage family, especially cauliflower. Cucumbers like to have other plants growing around the edges, such as beans or corn, which will protect them from the wind in a slightly enclosed space. Also, cucumbers grow well between annual "hedgerows" of corn. Stinging Nettle increases essential oil content of herbs.

(See GRAPEVINES, FRUIT TREES, MICROCLIMATE, *and* STINGING NETTLE*)*

IRON Stinging Nettle *(Urtica dioica)*, having an affinity for iron, is able to collect this element. Its mere presence is usually indicative of a good humus content, although it thrives on poor soil.

KNOTWEED *(Polygonum aviculare)* The polygonums comprise a large and common family found all over the world. This species, also known as Knotgrass, is often found along the paths in the garden. It likes to be walked on. Knotweed has a bad effect on turnips and is said to be troublesome to sheep. Since it grows only a few inches tall, it is easily overlooked by the gardener.

KOHLRABI *(Brassicaceae)* Kohlrabi, a member of the Cabbage Family, grows well with beets and onions, but has a harmful effect on tomatoes. It grows poorly near pole beans.

LAMB'S QUARTERS *(Chenopodium album, L.)* Lamb's Quarters in the potato field may be taken as an indication that the soil is tired of growing potatoes. It also indicates a well-fermented humus, a good soil. Makes very good eating, cooked as spinach.

LARKSPUR *(Delphinium consolida)* Larkspur is said to like winter wheat. There are hundreds of species of Delphinium, a number highly poisonous to cattle. Locoweed *(D. glaucum)* is the worst of these.

LAVENDER *(Lavendula officinalis)* Lavender herb will repel moths that attack woolen clothing and carpets. Lavender is difficult to grow in America. We might try growing it near a legume. On the Mediterranean slopes, it grows in association with Scotch Broom.

(See LEGUMES*)*

LEEKS *(Allium porrum)* Leeks (also onions) and celery do well together, sown in alternate rows. Composted pig manure is especially good for leeks and celery. Leeks do equally well in alternate rows with celeriac. The slender leek grows well between the bushy celeriac plants. Both are potassium lovers, and both do well on goat and pig manure. The leek is aided by carrots, and in turn, the leek helps repel the carrot fly from carrots.

(See CELERIAC, CELERY*)*

LEGUMES *(Leguminosae)* Legumes are plants which develop bacterial nodules on their roots, which bring nitrogenous compounds, available to plants, into the soil. The nitrogen component has been taken directly from the air. This benefits neighboring plants, and those following the legumes in the same soil. The leguminous roots promote growth in neighboring plant roots by giving nitrogen to the soil, and by aeration of the soil, which results from their own deep and luxuriantly growing roots. Among the smaller plants, the legumes include beans, peas, clover, alfalfa, lespedeza, peanuts, kudzu, and esparsette. Winter (hairy) vetch and other leguminous plants, such as soy beans and alfalfa, are used as green manuring – or cover – crops, because of their help as nitrogen fixers. Among the larger plants and trees in the legume family are:

> Kentucky Coffee Tree *(Gymnocladus dioicus)*
> Honey Locust *(Gleditsia triacanthos)*
> Judas Tree *(Cercis canadensis),* also called "Redbud"
> Yellow wood *(Cladrastis lutea)*
> Woad Waxen *(Genista tinctoria),* also called "Dyer's Greenweed"
> Scotch Broom *(Cytisus scoparius)*
> Black Locust *(Robinia pseudoacacia)*

Indigo Bush *(Amorpha fruticosa)*, also called "False Indigo"

Bristly Locust *(Robinia hispida)*, also called "Roseacacia"

Mesquite *(Prosopis juliflora)*, and there are many more.

As for clovers, red clover is the legume of soils — too acid, and too poorly aerated for alfalfa. It can bear a pH below 6.0, although the optimum is between 5.8 and 6.8. Every farmer knows well the value of legumes in his crop rotations, but they may be used in many other helpful ways, from the trees and shrubs of mixed hedges, windbreaks and mass plantings, down to the smallest garden plants. All of the plants in our gardens benefit by association with beans and peas. The lupines are often used as pioneers on poor soils, which they prefer. Dr. Robert F. Griggs tells us that lupines are the first seedlings to start growth on barren pumice after a volcanic eruption. He went to Alaska for a number of years, studying re-vegetation in the region of Katmai after its eruption in 1912. In its volcanic ash, almost devoid of nitrogenous compounds and microorganisms, the lupines were always the pioneers. Their roots showed abundant development of nitrogen nodules, as other plants followed close on their heels.

The legumes, with 15% mustard, help fruit trees and grapevines. *(See* CROP ROTATION, FRUIT TREES*)*

LEMON BALM *(Melissa officinalis)* Lemon Balm, or Melissa, as it is often called, is one of the fragrant herbs which radiates a beneficent atmosphere all around itself. As a tea by itself, either fresh or dried, or as an ingredient in a tea of mixed herbs, it is pleasant and refreshing.

Lemon Balm in pastures, is said to promote the flow of milk in cows.

Tea made of balm and Marjoram is given to cows after calving to strengthen and content them.

Perhaps the most valuable use we have found for Melissa is to rub the inside of the bee hive with a handful of the herb after hiving a new swarm. The swarm will never leave the hive. *(See* CAMOMILE, HONEY BEES*)*

LETTUCE *(Lactuca sativa)* Because lettuce is such a favorite, fast-growing vegetable, it has been observed for many years, and many facts are known about its preferences among companion plants. For instance:

Lettuce likes strawberries, is aided by the presence of carrots, and makes radishes tender in summer.

Lettuce in summer appreciates partial shade; early lettuce in very good soil aids onions. Biodynamic literature tells of successful combinations of lettuce, cabbage, and beets growing together.

LIGHT FEEDERS *(See* CROP ROTATION*)*

LILY OF THE VALLEY *(Convallaria Majalis)* If narcissus and lily of the valley flowers are put together in one bunch, they will soon wither. Similarly, mignonette increases the withering of flowers in a vase.

(See PRESERVATIVE INFLUENCE*)*

LOCUST, BLACK *(Robinia pseudoacacia)* Locust trees, being legumes, have a good effect on lima beans, making them grow unusually well. Plant locust as a border. Locust trees will help a dying forest. Lacking the trees, one might collect locust leaves for a special compost. In 1939, a scientist found toxic substances in the root excretions of *Robinia pseudoacacia*. Roots, leaves, and bark are poisonous if eaten. *(See* LEGUMES*)*

LOVAGE *(Levisticum officinale)* (*See* HERBS)

LUPINE *(Lupinus luteus* AND *Lupinus polyphyllus) L. luteus*, is the yellow, flowering bitter, and *L. polyphyllus* is very tall, with blue flowers. Both bitter and sweet lupine are used for reclaiming sandy soils. "A new piece of ground had just been bought. It was formerly a sand quarry and was just being brought under the biodynamic method. Sweet lupines were being grown on much of it as a preliminary measure. The whole garden was surrounded by thick hedges." This was written in 1937, pertaining to a garden in Europe, owned by Mr. Heinrich Thiess. He still operates it. Mr. Thiess became known as a very skilled observer of the manifold world of flowers and insects, vegetables, and animals living in the ground, which he tells about in many fine essays in German biodynamic literature.

Lupines leave behind them the finest, most friable soil.

(*See* LEGUMES)

MAPLE *(Acer)* The shallow, spreading root systems of most maple trees, are believed to excrete substances which inhibit many plants, especially wheat. The leaves have a remarkable preservative effect when layered in with apples, carrots, potatoes, and other root vegetables.

MARIGOLD *(Tagetes)* (*See* FRENCH MARIGOLD)

MARJORAM *(Majorana hortensis)* Marjoram is one of the most familiar of the aromatic herbs, easy to grow on almost any soil, and indispensable in every vegetable garden for its beneficial effect on surrounding plants. There are two varieties: Sweet Marjoram, *(M. hortensis)*, the annual, and Pot Marjoram *(Origanum vulgare)*, which is a perennial. Oregano is a close relative, very much like Sweet Marjoram.

Marjoram mixed in equal parts with Melissa (Lemon Balm) is used to make a healing tea.   *(See* HERBS*)*

MELON Melon leaves are extremely rich in calcium. They may be made into good compost. Germination of melon seeds is said to be stimulated by Morning Glory, *(Ipomoea)*.

MICE Leaves of the dwarf elder, *(Sambucus ebulus)*, will drive mice away from stored grain.

The vetchling, or everlasting pea, *(Lathyrus latifolius, L.)*, repels field mice and other small rodents.

The presence of Spearmint growing near a house is reported to repel various rodents. The Spurges, *(Euphorbia lathyrus and E. lactea)*, have the reputation of repelling moles and mice.

MICROCLIMATE is the climate from the surface of the soil up to the tops of the plants. In most field and garden crops, this is the space within seven feet above the ground. The temperature and moisture of this area are quite different from that of surrounding areas. Plants of infinitely varied habits and shapes and characteristics maybe used in various ways to create a microclimate. For instance, tall plants like sunflowers and sweet corn may be arranged to grow around low growing plants which like shade and limited circulation of air. Rows of corn can be used as windbreaks while a permanent hedge is getting established.   *(See* INTERCROPPING*)*

MILDEW, POWDERY *(Erysiphacea)* Powdery mildew on cucumbers covers both sides of the leaves with a white coat, out of which little black balls appear. This can be wiped off the outside of the leaf. Control the disease with Horsetail tea spray, mustard seed flour, or sulphur dust.   *(See* HORSETAIL*)*

MILK For increased milk supply in dairy cows and goats, feed members of the *Umbelliferae* family, such as Fennel, Caraway, and Dill, as a feed supplement.

Of recent years there is growing experimental evidence that milk as a spray can successfully be used to control virus diseases in tomatoes.

MINERAL ELEMENTS CONCENTRATED IN PLANTS Plants live in interaction with the surrounding soil. The presence, in excess, of individual elements can sometimes cause the plant to become toxic. On the other hand, a deficiency of these elements can sometimes be a blessing. One-sided fertilizing, for example, like high Potassium applications, can cause luxury consumption. The plant takes in more than it actually needs for optimum, metabolic functioning.

Plants are also active in enriching the topsoil with nutrient elements present only in small amounts in a particular soil. For example, the beech tree has been called "a pump for lime," because it enriches the topsoil with lime gathered at deeper levels, and returns it to the soil via the beech leaves. Lupines are pioneer plants on poor sands, because they also concentrate lime in their ash. Many leaves are valuable as compost material because they contain available trace elements.

MINT *(Mentha)* Spearmint and other mints, repels ants, and may help control aphis on nearby vegetation, since the ants place the aphis on the plants. Mint will also repel black flea beetles. Mint helps repel cabbageworm butterflies. Used indoors, Mint will repel clothes moths.

To keep flies off a milk cow, bruise some mint leaves in water, and give the cow a daily rubdown with it.

(See ANTS, GERMINATION*)*

MORNING-GLORY *(Ipomoea)* According to Golubinski, (1950), morning-glory stimulates the germination of melon seeds. *(See* MELON*)*

MOSQUITOES Mosquitoes are repelled by sassafras. The plant of the castor bean also repels them. Located near porches and patios, castor bean plants may increase the comfort of sitting outdoors on hot summer nights. If the castor bean were planted in quantities near marshy spots, it might even reduce the breeding of mosquitoes.

Fresh Pennyroyal leaves rubbed on the face and hands will protect one's skin from attacks of mosquitoes.

(See TANSY*)*

MOTHS, CLOTHES *(Tinea)* Moths that attack woolen clothing and carpets can be repelled by a sprinkling of dried leaves of Wormwood, Southernwood, Rosemary, Sage, Santolina, Lavender, and Mint. The same principle of using aromatic herbs as a moth repellent is used, for instance, in planting Tansy under a peach tree to repel fruit tree moths.

MULBERRY *(Morus alba* AND *M. rubra)* Mulberry is especially good for grapes, which can be trained to grow on the mulberry trees. Mulberry leaves can be used to repel worms in horses. Leaves are extensively used in silkworm culture. *(See* GRAPES*)*

MULCH A mulch of pine or spruce needles increases the vigor, flavor, and stem strength of strawberries.

Oak-leaf mulch or tan bark repels slugs, cutworms, and grubs of June bugs. Spread it on garden paths and in strips in the beds.

For further information about the general uses of mulch see garden books, especially those by Ruth Stout.

(See REFERENCES*)*

MUSTARD *(Brassica alba)* is a fast growing plant, used in Europe as a catch crop, and also as a cover crop for alfalfa. Fifteen percent mustard, sown with legumes, helps fruit trees and grapevines. Mustard and Rape – having somewhat alkaline root secretions – are ideal in a slightly acid or mineralized soil, such as one damaged, and in need of healing after the over-use of mineral fertilizer. The medicinal mustard plaster is made from its seeds. Field Mustard, *(B. kaber)*, and Black Mustard, *(B. nigra)*, deplete the soil of nutrients, and are regarded as a nuisance by American farmers.   *(See* HEDGE MUSTARD*)*

NASTURTIUM *(Tropaeoleum majus)* Sow to combat whitefly in the greenhouse. Nasturtiums planted near broccoli keep the latter free from aphis. Nasturtiums also repel squash bugs. When nasturtiums are planted under apple trees, they will keep away woolly aphis. They also aid radish. An elderly acquaintance in New England always planted a few nasturtiums among his potatoes. Nasturtium extract may be made according to the rules for making herb teas, and may be used as a spray on plants or trees, which would benefit from nasturtium influence.

(See HERBS, SPRAYS*)*

NEMATODES   *(See* FRENCH MARIGOLDS*)*

NETTLES   *(See* STINGING NETTLE, AFFORESTATION, COMPOST PLANTS*)*

NITROGEN The nitrogen-fixing bacteria in the soil are stimulated by the presence of Stinging Nettle. This may be one of the reasons for the beneficial effect of Stinging Nettle on all neighboring plants.

(See BUTTERCUP FAMILY, CLOVER*)*

NUT TREES Nut trees, especially the walnut, repel stable flies, houseflies, and flies that live on cattle. Nut trees in pastures are, therefore, a great comfort to cattle and horses. Such trees – if near manure piles and entrances to barn and stable – may reduce the number of flies.

(See CINQUEFOIL*)*

OAK *(Quercus)* The oak is hospitable to, and aids other trees. Citrus trees like the live oak as protection. Oak-leaf mulch or tan bark (from oak) repels slugs, cutworms and grubs of June bugs. Spread it on garden paths and along the garden beds between rows.

Biodynamic Preparation #505 is made from specially-treated, somewhat younger, red or white-oak bark, *(Q. rubra* OR *Q. alba)*. The oak tree has the special property of accumulating a tremendous amount of calcium in its bark during growth. In fact, the highest calcium ash content ever discovered, was founded in oak trees, growing on a sandy, calcium-deficient soil. The calcium content of this preparation is therefore very high, exceeding 10% by spectrographic analysis. This preparation in its effects stimulates the resistance of plants to disease.

(See COMPOST PLANTS,
MINERAL ELEMENTS CONCENTRATED IN PLANTS*)*

OATS *(Avena sativa)*, Oats and vetch, *(Vicia faba, L.)*, are good plant-affinities. Root excretions from oats have been found to inhibit the growth of young apricot trees.

ONIONS *(Allium cepa)* These bulbous plants belong to the lily family, along with leeks, garlic, shallots, and chives. They like to grow with beets. Onions in the garden are aided by Camomile, in the proportion of one Camomile plant to every four yards of onion row. In very good soil, early lettuce helps onions, as do a few plants of Summer Savory.

Onion rows alternating with carrot rows serve the double purpose of repelling both onion fly and carrot fly. Onions, garlic, and shallots inhibit the growth of peas and beans.

The Sea Onion, or Squill, *(Urginea maritima),* is native to southern Europe and Africa, where it is grown as a flower. The red variety is used as a rat poison.

Onion juice is effective on bee and wasp stings.

ORACH *(Atriplex hastata L.)* Orach checks the growth of potatoes. Whe Orach is harvested early, it can be an edible green, similar in flavor to spinach, which belongs to the same family. Another relative of the smooth pigweed is lamb's-quarters, *(Chenopodium album).*

(See LAMB'S QUARTERS*)*

OREGANO *(Origanum)* Oregano is one of the aromatic herbs. *(See* MARJORAM*)*

PANSY, WILD *(Viola tricolor)* Wild pansy, sometimes called Viola, will germinate nearly 100% if grown near rye. Without this influence, it is said to germinate only 20% to 30%. Conversely, rye is improved by a few pansies. This field pansy, common in Europe, germinates well when grown among rye, but is completely restricted when grown with wheat. Our cultivated pansies were developed from *V. tricolor*, which in turn, was derived from the violet.

PARSLEY *(Petroselinum hortense)* Parsley aids nearby roses, and is good for tomatoes. Parsley blooming in the garden is especially pleasing to the honeybees. There are five species in the genus, all natives of the old world.

PATCHOULI *(Pogostemon patchouly)* This is an East-Indian shrubby mint. The oil from it is used in perfumes. *(Research suggested)*

PEACH *(Prunus persica)* Soil from under old peach trees is toxic for young peach saplings. Curly leaf of peach trees may be cured by a mixture of *Equisetum* (Horsetail) tea, Stinging Nettle tea, pigeon manure, and Biodynamic Preparation #500. Mix and apply, at the rate of three gallons per tree, watered over the tree and its surrounding soil.

PEANUTS *(Arachis hypogaea)* The peanut originated in Brazil. Many horticultural varieties have been developed from it, so that it can be grown world-wide, even within various temperate zones. *(See* LEGUMES*)*

PEARS *(Pyras)* Root tip growth of pears was suppressed by the root excretions of grass. In other words, pear trees will grow better under open cultivation.

PEAS *(Pisum sativum)* Peas enjoy radishes, carrots, cucumbers, sweet corn, beans, and turnips. Peas are inhibited by onions, garlic, and shallots. Ordinarily, one does not grow peas two years in succession on the same ground. One writer states that peas, next to early potatoes, do not flourish, but most gardeners indicate that peas and potatoes do get along well, with two rows of peas alternating with early potatoes. The potatoes appreciate the nitrogen from the pea roots.

(See CARAWAY, LEGUMES, GERMINATION, GLADIOLUS*)*

PENNYROYAL *(Mentha pulegium)* This is a European, perennial mint. Pennyroyal repels ants. Fresh Pennyroyal leaves, rubbed on one's skin, form a protection against mosquitoes. American (mock) Pennyroyal, *(Hedeoma pulegioides)*, yields a commercial oil, which is used in soaps and medicines, and also to repel gnats and mosquitoes.

(See ANTS, MOSQUITOES*)*

PEPPERMINT *(Mentha piperita)* When peppermint is planted or strewn between cabbages, it protects the heads from the white cabbage butterfly. Peppermint growing with Camomile will be hindered in its oil production, while the Camomile itself benefits from this association, and will have a higher oil content.

Peppermint can be grown in a narrow bed, and can stand being walked on.

In one of Dr. Pfeiffer's experiments, Peppermint was grown with Stinging Nettle, to study the potential, symbiotic effects. Three rows of Peppermint alternated with one row of Stinging Nettle, which was then repeated. In the plants near the Stinging Nettle, the Peppermint oil content was 2.5%. In the control plants without the stinging nettle, the Peppermint oil content was only 0.75-1%. Peppermint oil has many commercial uses, and is extensively cultivated.

PHYTOPHTHORA BLIGHT *(Phytophthora infestans)* is a fungus that causes late blight of potatoes. When weather conditions are right, it multiplies with great rapidity, and turns the vines black, almost overnight. This is the organism that caused the Irish potato famine of the 1840s, and is still a threat.

(See APPLE TREES, POTATOES*)*

PINE *(Pinus)* A mulch of pine needles increases the vigor, flavor, and stem strength of strawberries.

Pine suppresses the growth of wheat. Rain, washing down over pine needles, contains secretions which have a deleterious effect upon the germination of seeds. This could be experimentally confirmed, by using water extracts from pine needle litter. Pine branches, decomposing in soil, noticeably inhibited growth of bacteria and germination of seeds – more so in an acid medium (pH 5.6), than in a neutral one.

(See FOXGLOVE*)*

PLANTAIN *(Plantago)* Dr. Pfeiffer observed that plantain and red clover are frequently associated. There are some 200 species of plantain recognized.

POLLINATION Summer flowering plants of mixed varieties attract a wide and balanced variety of insects, thus promoting pollination of all neighboring vegetation.

POPPY *(Papaver)* The field, or Shirley poppy, *(P. rhoeas)*, sometimes infests grain fields, and has such an effect, especially on winter wheat, that the grain crop is limited to a small amount of lightweight seed.

Poppies are robbers of the soil. They are sometimes grown for seed and oil, but the soil needs quite a bit of rest and reinforcement afterwards. Dutch farmers grew a crop of poppies in order to choke out weeds, which they could not get rid of by any other means. In gardens, poppies are grown for the beauty of their many colored blossoms. The seeds can lie dormant in the ground for years, and then show up again with a grain crop.

Poppies dislike barley, but will lie dormant until winter wheat is sown in the field.

POTASSIUM LOVERS Leeks and celeriac love potassium. They appreciate goat and pig manure.

Tobacco accumulates potassium; hence, if composted, its stalks and unused leaves should provide enrichment of that element. One possible problem in composting modern tobaccos in the United States, is that they may have be so heavily sprayed with chemicals, that they will not compost well.

Other plants that are rich in potassium, which may be incorporated into the compost heap are: Tansy, Russian Thistle *(Salsola kali)*, and all of the thistles and sunflowers. Wormwood indicates the presence of potassium in the soil. *(See* CELERIAC, WOOD ASHES*)*

POTATO *(Solanum tuberosum)* Early potatoes are known to grow well with beans, sweet corn, cabbage, and peas. Plant two rows of peas, alternating with early potatoes. The potatoes appreciate the nitrogen from the pea roots.

Potatoes and English broad beans are good companions. The broad beans need air and grow better if planted not more than three rows thick. Potatoes following a rye crop are observed to grow exceedingly well.

A small amount of horse-radish, one plant in each corner of the potato patch, aids the general health of potato plants. Dead Nettle, Sainfoin *(Esparsette),* and nasturtium, according to old-time New Englanders, benefit potatoes when growing nearby.

If hemp grows in the neighborhood of potatoes, they are not likely to be attacked by *Phytophthora infestans*, the cause of late-blight. It has also been observed by scientists, that resistance of potatoes to late-blight is lowered in the vicinity of sunflower, tomato, apple, cherry, raspberry, pumpkin, or cucumber. Potatoes grown in the vicinity of a birch wood, rot more easily than potatoes grown in the vicinity of pine. *(See* GARLIC*)*

Potatoes and sunflower stunt each other.

Root excretions from potatoes somewhat inhibit growth of tomatoes.

Potatoes should not be grown near Orach, because the latter checks growth of the former. Orach is a relative of pigweed. It prospers well next to the potato, but if it increases, it indicates that the soil in the potato field is exhausted. *(See* ORACH*)*

Plant early potatoes in the garden, and after the first hilling with a hoe, plant cabbages between them. The cabbages should do well.

Potatoes growing much foliage suppress many weeds. If the soil becomes infested with lamb's quarters, it is a sign that potatoes have been grown there for too long, and it is time to change crops.

Rows of potatoes between bean rows repel the Mexican bean beetle.

Rows of green beans, (any, but lima beans), alternating with rows of potatoes, will keep away Colorado potato beetles. Flax, in the rows between potatoes, remarkably reduces the number of potato bugs. Research is presently being done to determine the relationship of the marigold to the potato eelworm, or nematode. There is also some evidence that shamrocks imported into the United States from Ireland in 1964 have had to be quarantined, to check against importation of the golden nematode, "A small, worm-like pest — which does serious damage to potato and tomato crops — but does not harm shamrocks, and only hitches a ride in the soil, clinging to its roots." It is to be hoped that the marigold will control these nematodes, would they become established in this country.

Colorado potato beetles prefer eggplant to potatoes. A border of eggplants around a potato patch gives one a chance to concentrate, catch, and destroy a large portion of the attacking potato beetles.

POTATO BUGS Colorado potato beetle, *(Leptinotarsa decemlineata),* and the so-called, old-fashioned potato bug, or Blister beetle, *(Epicauta vittata),* can both be kept under control, by the use of certain, common plants which repel them. *(See* EGGPLANT, FLAX, FRENCH MARIGOLDS, INSECT PESTS, MARIGOLDS, POTATOES*)*

PRESERVATIVE INFLUENCE OF NEIGHBORING PLANTS Certain plants are known to have a preservative effect on neighboring plants after they are harvested or cut. Purple foxglove, lily of the valley, and plants of the scilla genus, are among those known for this effect. Another plant with this effect is the Stinging Nettle. Tomatoes grown near it, are known to keep better.

(See STINGING NETTLE, SUGAR MAPLE*)*

PUMPKIN *(Cucurbita pepo)* Pumpkins and corn are good neighbors, but pumpkins and potatoes do not like each other. *(See* CORN, POTATOES*)*

PYRETHRUM *(Chrysanthemum cinerariaefolium)* In biodynamic literature, there is an account of a farm visit in Germany in 1937. "We had time to walk through their gardens where... the association of plants is carefully studied. Pyrethrum was planted alongside strawberries as a pest preventative."

The flowers of this plant are the source of the spray and dust known as Pyrethrum and generally considered a safe pesticide.

QUACK GRASS *(Agropyron repens* OR *Triticum repens)* can be choked out by sowing soybeans, cowpeas, and millet, if the land was first thoroughly cultivated, and the weather was hot and dry. It can also be choked out by two successive crops of rye. Best control is by a heavy mulch, which starves out the underground parts.

In lawns, quack grass can be controlled by hand pulling before it forms a mat. It takes frequent cultivation in the garden, and much patience. Dry, hot weather will wilt the roots if they are brought to the surface.

Tomatoes will gradually choke out quack grass; they have an antipathy for this hard-to-get-rid-of weed.

RABBITS *(Oryctolagus cuniculus)* Rabbit manure – liquefied and painted on tree trunks – is reported to prevent the cottontail, *(Sylvilagus floridanus),* from girdling or injuring such trees. "Rabbits will not go through a wall of onions." Any of the onion family, even ornamental Alliums, will protect a garden, if planted as a border around it. This is a simple, applicable principle which might be more widely practiced.

RADISH *(Raphanus sativus)* Radishes, both black and white, aid other vegetables. Peas and radishes are mutually helpful. Radish growing near nasturtium is aided, and has "a perfect flavor." Leaf lettuce in summer makes radishes tender. Chervil and radishes in alternate rows are mutually helpful, and the Chervil makes the radishes taste hot. Pole beans like radishes.

A few radish seeds sown beside cucumber plants or other vine crops, will help keep away striped cucumber beetles *(Diabrotica vittata).*

Radish and Hyssop dislike each other. *(See* CARROT FLY*)*

RAPE *(Brassica napus)* Rape loosens heavy soil and leaves it friable, and improves drainage because of its deep tap-root. It also helps soil which has been injured by over-doses of mineral fertilizer.

If Rape is grown near Hedge Mustard and Field Mustard, it will be inhibited in growth.

RASPBERRY *(Rubus)* Keep raspberries away from potatoes, as the latter are then more susceptible to blight. Raspberries and blackberries should not be grown together.

ROOT GROWTH When scientists of the future have investigated the dynamic happenings in the root areas, as thoroughly as they have studied the chemical composition of substances of the same plants, we will know much more about plant symbiosis. We already know that every plant has an extensive root system below ground, which reaches astronomical proportions when measured or weighed. "Dittmer has estimated, that a single rye plant, under favorable conditions, develops as a seasonal total, a root surface, including root hairs, of 6,875 square feet, and a total root length of 387 miles, carrying 6,603 miles of root hairs. The average daily increase of root lengths alone, was estimated at more than three miles." This complex root system growing invisibly below ground, may in its living processes, create substances, or dynamic effects, which influence neighboring plants simultaneously.

Even after the removal of what we think of as the "plant," much of the root network remains in the soil, where it exerts influences on the plants of the next generation, either of the same species, or of others. Roots and stubbles are the residues left by plants in the field. These are actually a second crop. There are some figures on them that vary from place to place, and from year to year. The following are average figures:

SMALL GARDEN	800-1500 LBS. PER ACRE
PEAS, BEANS ETC.	400-2000 LBS. PER ACRE
POTATOES, BEETS, ETC.	500- 900 LBS. PER ACRE
RED CLOVER, GRASS, 2-YEARS	2400-7700 LBS. PER ACRE
ALFALFA, 4-YEARS	4500-5500 LBS. PER ACRE

These residues become food for the soil organisms. They are raw material for the formation of stable humus. By their decomposition, they release plant nutrients, and contain a multitude of growth-enhancing or inhibiting substances. The biological activity of the soil is maintained by crop residues, manures, and composts.

ROSEMARY *(Rosmarinus officinalis)* Rosemary and Sage have a stimulating effect upon one another. Rosemary repels carrot worm butterfly. Rosemary is a favorite among the aromatic herbs, being one which overwinters successfully in the house, and increases every year in size.

(See HERBS*)*

ROSES *(Rosa)* Roses and garlic are mutually helpful, as has been proved by innumerable, American rose gardeners in the past few years. From Europe, we learn that Bulgarian gardeners grow onions and garlic as intercrops with roses, and it has been observed that the roses with the garlic, produce the stronger perfume in larger quantities, than without garlic. This is one of the great mysteries of companion planting. It is even helpful if one fertilizes rose bushes with compost made with garlic and onion refuse. The onions also repel rose bugs.

Roses are also aided by the nearby presence of parsley. Mignonette is another companion plant which roses like. It makes a low ground cover and border plant for rose beds, and at the same time, benefits the roses. Another companion plant favored by roses is lupine, particularly the perennial lupines which help increase the soil nitrogen and attract earthworms.

One plant the rose does not like is boxwood, which has out-spreading, woody roots, that interfere with the roots of the rose bush. It is better to plant roses with plants whose roots go deep, rather than widespread.

RUBBER TREE *(Hevea)* Citrus trees appreciate the protection of the rubber tree.

RUE *(Ruta graveolens)* Rue and Sweet Basil dislike each other.

Rue repels houseflies and stable flies. Grown in window boxes or around manure piles, or as a border around barns, this herb should help to exclude these pests from barns and houses.

RYE *(Secale cereale)* Rye is aided by a few pansies *(viola)*. Rye is also considered a good weed fighter. It inhibits the germination and growth of the field poppy, and when planted twice in succession, can eliminate quackgrass. Rye will also fight such low growing weeds as chickweed, which survives winter. Further study has shown that rye suppresses the growth of weeds, more so than wheat, as it hinders germination of the seeds and seedlings growing near it. The weeds either die, or their growth is checked.

(See PANSY*)*

SAGE *(Salvia officinalis)* Sage and Rosemary stimulate one another. If Sage is grown among cabbages, it protects them against the cabbage butterfly. Even twigs of sage strewn across the cabbages will have a beneficial effect of repelling this pest. Sage also aids cabbage, making the cabbage plants more tender and digestible. It aids many other vegetables in general.

The helpful effects inherent in the Sage plant can be utilized in Sage tea, made with boiling water, to extract the herbal qualities. This tea should not be applied to young plants, since it might retard their growth, but it is safe to use on older ones, which have passed the stage of blooming.

A report in *Science Newsletter* 85:151, March 7, 1964, tells of research determining adverse effects of some indigenous members of the Salvia Family – "The tiny roots of a young cucumber seed stop growing, when they come in contact with a volatile chemical produced from aromatic shrubs. Grass seedlings also seem to be affected by this material, produced in the leaves of the shrubs *Salvia leucophylla, S. apiana,* and *Artemisia californica.* In grassland fields in the Santa Inez Valley of California, bare soil stretches under and beyond the canopy of the shrub branches. Evaporation from uninjured shrub leaves is deposited or trapped in dew, which comes in contact with the seedlings, and stops them from growing," reported biologists, Cornelius H. Muller, Walter H. Muller, and Bruce L. Haines of the University of California, Santa Barbara.   *(See* WORMWOOD*)*

SAINFOIN *(Onobrychis viciaefolia)*   *(See* ESPARSETTE*)*

SALSIFY, OYSTER PLANT *(Tragopogon porrifolius)* A variety of salsify called Black Salsify *(Scorzonera hispanica)*, helps carrots by repelling the carrot fly, even more effectively than do onions. *(See* CARROTS, ONIONS*)*

SANTOLINA *(Chamaecyparis)* This is a small genus of Mediterranean plants belonging to the thistle family. It looks a little bit like yarrow. Santolina is a moth repellent.

SAPONIN The saponin-rich plants now appear to work beneficially on follow crops. Saponin is a soap-like, lather producing substance occurring in many different plant families. The humifying remains of these plants provide an environment favored by successional plants. It is an effect in time, rather than in space. Important saponin containing plants (unrelated) are the primroses, some carnations, many legumes, labiates, horse chestnut, camellia, spinach, orach, pokeweed, cyclamen tubers, runner beans, tomatoes, potatoes, the violas, and mullein. Probably the most familiar saponin-rich plant is Bouncing Bet or Soapwort *(Saponaria officinalis),* used even today, for cleansing very delicate fabrics which might be injured by soap.

SASSAFRAS *(Sassafras albidum)* Sassafras is said to repel mosquitoes. The bark from young roots makes a delicious tea.

SAVORY, SUMMER *(Satureia hortensis)* Summer Savory as a border around onions helps them. It also grows well, and is beneficial when planted near green beans. Use this herb to cook with them, as well. *(See* HONEYBEES*)*

SAVORY, WINTER *(Satureia montana)* Winter Savory is an insect repellent. It also inhibits germination of certain seeds. *(See* GERMINATION*)*

SAWDUST MULCH Fresh sawdust mulch, from coniferous trees only, has been tried and proven effective in protecting carrots from the carrot fly. Sawdust and wood shavings, being acidifying materials, are widely used as a mulch for blueberries. For ordinary crops, this mulch should be composted, or supplemented with plenty of manure before it is worked into the ground.

SCOTCH BROOM *(Cytisus scoparius)* Scotch broom is a legume which accumulates calcium. If it is kept in check it aids the soil. *(See* CALCIUM, LEGUMES*)*

SEA ONION, SQUILL *(Urginea maritima)* is good for colic in cattle, but must be used judiciously, because it can also act as a poison. It is one of the oldest Egyptian medicine plants, and is still used for making a variety of cathartics and emetics. *(See* ONIONS*)*

SESAME *(Sesamum orientale)* is highly sensitive to root excretions of sorghum, *(Andropogon sorghum)*. Fletcher observed that sesame cannot ripen in the vicinity of sorghum.

SHADE LOVERS Certain plants appreciate partial shade during the hot summer. Some of these are lettuce, Chervil, melons, squash, and pumpkins – which all do well in the cornfield because of its protective shade.

SHALLOTS *(Allium ascalonicum)* Shallots, onion, and garlic inhibit the growth of peas and beans.

SILICA One plant that is especially rich in silica is the prostrate Knotweed, *(Polygonum aviculare)*, which is very common all over the world, growing on garden borders, and along paths.

"All equisetum plants, the common Horsetail *(Equisetum arvense)*, in particular, have a silica skeleton. Burning the green stems and leaves in a hot but quiet flame, removes all organic parts, but leaves a white skeleton of silica, which will show the original structure of the little stems." It is this silica content (as high as 80% of the entire ash), which made the Horsetail a valuable tea remedy to our ancestors. A 0.5-2.0% solution, made by boiling for 15-20 minutes — is applied as a biological spray against mildew and other fungi — on grapevines, vegetables, roses, and fruit trees. This tea, while not as powerful as copper and arsenic sprays, has a gentle and swift action, which does not disturb the soil life. However, the first cause of fungus infection, too moist a stand of the infected plant, or too rich fertilizing, should be corrected before the Equisetum spray is used.

Quack grass and Stinging Nettle are also rich in silicic acid. *(See* HORSETAIL*)*

SLUGS *(Deroceras agreste* AND *D. gracile)* When garden soil has become rich and fertile, and capable of bearing succulent vegetables, the slugs thrive, especially in wet seasons. The slug is related to the snails, but has only a thin plate for a shell. This creature is called a gastropod, because its stomach is in its thick, slimy "foot," which attaches itself to green leaves and proceeds to devour them. Slugs have a sense of smell, and travel in the garden at night, leaving a narrow trail of slime as they creep from one plant to another. They can climb to the top of any plant to eat the choicest leaves. They are particularly destructive of lettuce and cabbage heads.

Slugs may be repelled by a mulch of tan bark or oak leaves in the garden. Wormwood tea sprinkled over the soil in spring and fall repels them by its bitterness. Wood ashes placed around the plants cause slugs to secrete mucus to the point of exhaustion and death.

(See MULCH, WORMWOOD*)*

Slugs will creep under grapefruit shells or even flat boards set out in the garden. After they have gathered under the bait, they can be picked up and destroyed. Salt sprinkled from a saltshaker directly on the slug will cause it to shrivel, lose vital moisture and ultimately die. Quicklime sprinkled around the garden area will keep slugs away. It has to be renewed after rain. Slugs are attracted to shallow saucers of beer. They drink the beer and drown.

SOLUTION To make a two percent solution (2%), take any amount in weight, and add to it 50x that amount of water. 100x the amount would make a 1% solution, and so on.

SORGHUM *(Andropogon sorghum) (S. vulgare)* Sorghum is toxic to sesame and to wheat.   *(See* SESAME, WHEAT*)*

SOUTHERNWOOD *(Artemisia abrotanum)* Southernwood, a European wormwood, is a moth repellent, and will protect nearby cabbages against the cabbageworm butterfly. Plant southernwood near fruit trees to repel fruit tree moths. *Artemisia absinthium,* introduced here, might well be tried in the same capacity.

SOYBEAN — SOYA BEAN *(Glycine max, G. soja)* Soybeans will loosen heavy soil and leave it friable. They will also choke out weeds, as they grow faster and thicker. They enrich poor soil, as do other legumes. It is native to China, where it has been cultivated as it's principal crop for around 5,000 years.

SPEARMINT *(Mentha spicata)* Spearmint is reported to repel various rodents. It repels ants, and may help to control aphis on nearby vegetables, since the ants place the aphis on the plants. *(See* APHIS, MICE*)*

SPINACH *(Spinacia oleracea)* Spinach likes strawberries.

(See INTERCROPPING*)*

SPRAYS In using biodynamic sprays made of herb teas – such as Equisetum, Sage, Wormwood, Stinging Nettle, Nasturtium extract, Camomile, and others – it is advisable to add a very small amount of diluted clay as a spreader, especially for tomatoes, vine crops, and the like.

SPRUCE *(Picea)* Spruce is aggressive and hostile to other trees. It is said, that the adverse effect of spruce may be noticed in the soil for 15-years after it is cut off. However, a mulch of spruce needles increases the vigor, flavor, and stem strength of strawberries. The latter like to grow near spruce trees.

SPURGE *(Euphorbia lathyrus* AND *E. lactea)* These spurges are called Mole Plants, and will repel moles and mice.

(See EUPHORBIA*)*

SQUASH BUG *(Anasta tristis)* Squash bugs are repelled by nasturtiums.

SQUILL *(Urginea maritima)* *(See* SEA ONION*)*

STINGING NETTLE *(Urtica dioica)* Stinging Nettle has at least three properties which illustrate its dynamic character:
1) It helps neighboring plants to grow more resistant to spoiling. *(See* TOMATOES*)*
2) It changes the chemical process in neighboring crops. F. Lippert, who for a long time managed the herb garden of a drug company, reports that Stinging Nettle, growing as a companion plant, increased the content of

essential oil up to 20% in Valerian, over 80% in *Angelica archangelica*, 10-20% in Marjoram, 10% in Sage, and 10% in Peppermint. *(See* PEPPERMINT*)*

3) It stimulates humus formation. This third property can be studied if one digs out the soil close to a nettle root and observes the kind of humus that is formed there — a blackish brown neutral humus. The leaves and stem of this plant rot to an ideal humus. There may also be certain secretions on the roots which stimulate life and fermentation. The biodynamic method of gardening uses a humus made from Stinging Nettle in order to stimulate fermentation in the compost or manure pile.

The Stinging Nettle bears fine hairs on the leaves and stems, which contain formic acid, and perhaps a yet unknown poison, (which may be neutralized by rubbing the skin with jewelweed, *Impatiens biflora* OR *I. pallida*, or with any member of the sorrel family, including rhubarb). Nettles were known in oldest times for their medicinal value. They increase blood circulation and act as a stimulant. The very young plants cook like spinach in early spring, and are an excellent table green. Baby chicks devour chopped nettle leaves greedily, as nettle is both nourishing, and a preventive of coccidiosis and/or diarrhea in poultry.

Stinging Nettle is rich in vitamins and iron, which was recognized in olden times, when it was used as a remedy against anemia, and to strengthen the vitality of those who ate it.

To make the fermented extract, which is a liquid herb manure, cover the cut herb with water, and allow it to decompose for three weeks. The nettle plants will be completely digested. During the three weeks and after, this nettle liquid will promote plant growth, and protect plants against unhealthy conditions. It is sprayed on the plant to strengthen it, to help it overcome drought, etc. *(See* BLACK APHIDS, COCCIDIOSIS, HERBS, IRON, PEPPERMINT, TOMATOES*)*

STRAWBERRIES *(Fragaria)* In biodynamic practice, cultivated strawberries have often been accompanied by their favorite companion plants – bush beans, lettuce, spinach, and especially Borage. Known and recorded many years ago, Pyrethrum was planted alongside strawberries as a pest preventative. They also do well near a spruce hedge. Strawberries dislike cabbage.

Strawberries like compost containing pine needles and straw, and they also appreciate a mulch of pine needles or spruce needles. It is said that the pine needle mulch makes the berries taste more like wild strawberries. A special compost for strawberries is made of straw, pine needles, some green stuff with the usual inter-layers of soil and a small amount of lime, and the Biodynamic Preparations 502 and 507.

One writer suggests planting legumes as intercrops and a little Thyme as a border plant.

F. C. King recommends that soil in which strawberries are grown be left undisturbed, and that resinous sawdust mulch be applied in October, rather than in spring, when it is apt to dry out and cool the surface of the soil, thus inviting frost damage to flowers.

SUDAN GRASS *(Sorghum vulgare sudanense)* A slender annual grass, introduced from Africa widely cultivated in semi-arid regions, used for hay. Where moisture is adequate, it is often grown with soybeans.

SUNFLOWERS *(Helianthus annuus)* Sunflowers and potatoes stunt each other.

Sunflowers are heavy feeders, and require a generous amount of good compost. They grow so tall and sturdy, that they make a good border planting, to protect the edges of the garden from wind, and from undesirable visibility. The seeds are most attractive to birds, and will insure a good bird population, as long as there are seeds in sight. Sunflowers in bloom also give the bees a pleasant source of pollen and nectar.

Research from 1947 indicates that potatoes in the vicinity of sunflowers were likely to be infected with *phytophthora* blight.

Cucumber grows well near sunflower, which provides a satisfactory windbreak for it and other vine-crops.

SWEET BASIL *(Ocimum basilicum)* Rue and Sweet Basil are harmful to each other, and will not grow near each other in the same garden bed.

SWEET CLOVER *(Melilotus)* "The (three) species grown in this country contain coumarin, a substance having a vanilla-like odor. Spoiled sweet clover hay, and poorly preserved sweet clover silage is frequently toxic to animals, causing both external and internal bleeding. The basic principle which causes this toxicity is a decomposition product of die coumarin, which develops as the sweet clover decomposes."

(Grass: The Yearbook of Agriculture, 1948)

Coumarin also exercises an inhibitory effect on seed germination, and growth of top and roots of corn, as reported in recent literature on experiments in agronomy.

SYMBIOSIS is a Greek word meaning, *living together*, and today's scientists' name for the phenomenon which observant gardeners and farmers see happening among their plants, and also between plants and animals every day during the growing season is, *a living together of dissimilar organisms in such a way, that their mutual influence is beneficial to each of them.* Antibiosis is the contrary of symbiosis, as is antipathetic symbiosis.

TANBARK (the spent bark which has been used in the tanning process, which has a very bitter principle remaining), used as a mulch on garden paths, and in strips along the beds, will repel slugs, cutworms, and June bug grubs.   *(See* OAK*)*

TANSY *(Tanacetum vulgare)* Tansy repels flies and ants. It is also a moth repellent. Planted beside peach trees, it will keep away harmful, flying insects.

An introduced plant, Tansy has been called "Parsley Fern" from the shape of its leaves. It grows in waste places, and has a very strong, bitter, aromatic odor. The blossoms, when dried, do not wilt, hence the Greek name *Tanacetum*, which indicates immortality. Distilled Tanacetin Oil from Tansy, has been used as a fly and insect repellent. In medicine, Tansy has been recommended against intestinal worms, and in wine against stomach and intestinal spasms. The Russians have used it as a substitute for hops in beer. Both stems and leaves are poisonous to livestock and man. Rubbed on the surface of raw meat, it will protect it from flies. Rubbing it into a dog's fur helps repel fleas in late summer. Tansy concentrates plenty of potassium and, therefore, may well be used in the compost heap.

TARRAGON *(Artemisia dracunculus)* Tarragon is one of the perennial, aromatic herbs, and belongs to the same family as the wormwoods. It belongs somewhere in every Biodynamic garden. It likes light, well-drained soil. Tarragon-flavored vinegar goes well in salads.

THISTLE, BLESSED *(Cnicus benedictus)* This is a valuable thistle-like plant with medicinal and industrial uses. It is also used as a basic ingredient in Benedictine liqueur and other bitter tonics (stomach bitters). Blessed thistle is said to repel certain insects but more experiments are needed. *(See* HERBS*)*

THISTLE FAMILY *(See* POTASSIUM LOVERS, WHEAT*)*

THORN APPLE *(Datura stramonium)* The thorn apple, or Jimson Weed, accumulates phosphoric acid, and its leaves may make a compost richer in phosphorus. The plant is poisonous to domestic animals and human beings.

THYME *(Thymus vulgaris)* Thyme helps repel cabbage worms and by its aromatic qualities enlivens the plants nearby in the garden. Many like it as a kitchen herb.
(See GERMINATION, HONEY BEES*)*

TOADS *(Bufo)* Toads are very useful in the garden, eating many insects and pests which attack our plants. They exude a slime distasteful to enemies, but not poisonous to man. Their skins contain adrenalin, and have been used medicinally. Toads do not cause warts.

TOBACCO *(Nicotiana tabacum)* Tobacco, like the tomato, likes to grow in the same place year after year. Like the tomato, it also likes to be fertilized with compost made of its own leaves. Tobacco accumulates potassium. Tobacco juice is effective on bee and wasp stings.

TOMATOES *(Lycopersicon esculentum)* Tomatoes and asparagus are mutually helpful. Parsley is a good companion to both. Tomatoes aid early cabbage. Tomatoes and brassicas of all varieties grown together will help to ward off the white cabbage butterfly.

Tomatoes give off root excretions, which have an inhibitory effect on the growth of young apricot trees. It was also found that potatoes grown near tomatoes were not resistant to potato blight.

In 1950, Porozhikov noted that certain, volatile substances of tomatoes had an inhibiting effect on certain insects attacking the gooseberry shrub. According to this author's observations, gooseberry shrubs planted near tomatoes do not suffer from these insects.

Tomatoes like to grow in the same area year after year, and prefer compost made with their own stalks and leaves.

They do not grow well in the vicinity of kohlrabi and Fennel.

Tomatoes are aided by Stinging Nettle. The presence of the latter growing nearby makes the tomatoes keep better, with little mold or putrefaction.

(See GARLIC, MARIGOLD, SPRAYS*)*

TOXIC SUBSTANCES occurring in soils as products of root excretions, plant residues, and microbiological metabolism, have recently been studied most intensively. This research opens new avenues towards an understanding of plant relationships, the necessity for rotation, intercropping, etc. Actually, growth-enhancing and inhibiting, germination-enhancing and inhibiting factors are all side-by-side in the same soil. One-sided cropping and poor management shifts the equilibrium to one side. Krasil'nikov, who gathered not only the world's literature about this subject, carrying out many investigations himself, demonstrated the occurrence of toxic substances, founded mainly in the naturally wooded region around Moscow, where soils are considerably leached.

The toxicity of soils was determined by the viability of azotobacter bacteria, and by the germination of certain seeds including wheat and beets. At the same time, a total count of the microflora was made, including the microbial inhibitors that form toxic substances.

These studies have shown that many soils contain toxic substances. In forest soils, as a rule, there are more of these inhibitors than in forest-free soils; while there are less in plowed soils, and still less in well-cultivated ones.

The toxicity of forest soils is determined by the varieties of trees growing there. The greatest amount of toxic substances were found under spruce groves, and in lesser amount under pine and aspen groves. Soils under birch and oak were weakly toxic or entirely non-toxic.

TULIP *(Tulipa)* Tulips suppress the growth of wheat.

TURNIP-RUTABAGA *(Brassica rapa* AND *Brassica napobrassica)* Turnips and peas are mutually beneficial. Turnips are harmed by hedge mustard and by knotweed.

(See MUSTARD, KNOTWEED*)*

In 1964, an insecticide was isolated from the roots of turnips and other cruciferous crops. It is called 2-Phenyl-ethylisothiocyanate, and was tested for its lethal effect on vinegar flies and house flies, under laboratory conditions.

The substance is contained in turnip peelings, and in root tissues of most other members of the cruciferae family.

VALERIAN *(Valeriana officinalis)* In a border, Valerian helps most vegetables. It is a specialist for phosphorus — that is — it stimulates the phosphorus activity in its own vicinity. Valerian attracts earthworms. (N.B. it also attracts cats if the roots are bruised.)

The influence of the Valerian plant can be enhanced by making a spray of the juice. Of Valerian spray it has been written that it is, "a particular joy for the earthworms, and they are attracted by it. The Valerian spray is repeated once a month during the summer, and it encourages the general health and resistance of the plants. It can be sprayed on the soil, also on all plants at whatever stage, whereas caution is necessary with some other sprays."

VETCH *(Vicia)* Vetch is often grown as a companion crop for the cereal grasses, especially rye, to their benefit. A mixture of rye and hairy vetch *(V. villosa),* when planted in the fall, is one of the most valuable green-manure crops. Vetch is often sown in orchards. Being a legume, it enriches any soil with nitrogen and humus.

VETCHLING *(Lathyrus pratensis, L.)* The vetchling, or perennial sweet pea, repels field mice (the European vole). The ripening seeds of this plant are poisonous to grazing animals, also to people.

WALLFLOWER *(Cheiranthus cheiri)* Wallflowers have been reported to be good for apple trees.

WALNUT *(Juglans nigra, J. cinerea, J. regia)* The Black Walnut, the Butternut, and the English Walnut have an inhibiting effect upon the growth of tomatoes, potatoes, and other plants. *(See* NUT TREES*)*

WEEDS Many of the plants we have spoken of as friendly neighbors for our garden vegetables and field crops, are known to many people as "weeds." They root them out of their gardens and fields, and nowadays, kill them with powerful sprays.

There are many people who speak of the value of these wild plants in the life-cycles of nature. Chief among these in America is Joseph A. Cocannouer.

Out of long years of observation and field experiments, both here in this country, and abroad – in the Philippines, China, Africa, Mexico, and Europe – he has since put down his thoughts in a book, *Weeds, Guardians of the Soil*, (New York, Devin-Adair).

He urges the controlled use of weeds as companions to field and garden crops. The use of deep-rooting ones helps to bring back eroded, compacted soils. These, such as pigweed *(Amaranthus retroflexus)*, not only bring up nutrients from deeper soil layers, but provide what he calls "fibre" for the soil, as they help to build up the "sponge" structure of the surface soil, as well as the sub-soil. This pigweed is described by Professor Cocannouer as a valuable "mother plant" with potatoes.

He also points to the value of weed and legume mixtures to aid in bringing back worn out pastures.

Nor does he neglect the value of weeds in the compost heap – the greater the variety in the mixture, the better.

One of the most striking examples cited by Professor Cocannouer, is that of an orchard in California. The owner pushed this orchard to higher and higher yields with fertilizer, to profit from the good market conditions. Then the trees began to look weak and unthrifty. Experts said he must start a new orchard on new land. He let the old orchard go to weeds – doing nothing but to irrigate it, occasionally, during the season. To his astonishment, the orchard began to revive, especially where the weeds were thickest. He had, undoubtedly, continued his new method of cultivation, using weeds as the "mother crop" in his orchard.   *(See* COMPOST*)*

WHEAT *(Triticum vulgare)* Wheat and maize are good companions, and the growth of the former is increased when they are sown together.

Wheat is aided by Camomile *(Matricaria chamomilla)*, if in a very small ratio (not over 1:100), but harmed by a large amount of camomile in field crops.

According to recent experiments, the growth of wheat is suppressed by dogwood, cherry, tulip, and pine. Another report states that wheat is adversely affected by the roots of sorghum.

The growth of wheat is checked by poppies.

The Agricultural Experiment Station in North Dakota reported in 1950 that aqueous extracts of field bindweed *(Convolvulus)*, and Canada thistle *(Cirsium arvense)*, are harmful to wheat and linseed.   *(See* GERMINATION*)*

WHITE FLY *(Trialeurodes vaporariorum)* White fly in tomatoes is reported to have been eliminated by French marigold.

A story is told, of a grower who rented some old greenhouses that were infested with "crowds of white flies..." In the spring, he raised nasturtium plants in seed boxes. When he set out his tomatoes (which are especially susceptible to white flies), he interspersed them with nasturtium. "After a time, the nasturtiums had to be cut back a little, so that they would not cast too much shadow on the tomatoes. It was certainly successful. You had to look for a long time before finding any white flies. Only when the tomatoes were fully ripe were there a few more, but it was tolerable.

"The white fly belongs to the same class as the cochineal insects and aphis, the woolly aphis, that most terrible pest of apple-trees, being one of them. At the end of the Agricultural Course (of Dr. Rudolf Steiner), we read the sentence, 'Under the trees that suffer from woolly aphis, a ring of nasturtium should be planted.'

"Nasturtiums have a strong, aromatic essence which passes through the roots into the surrounding soil. The tomato roots take it up into their sap, and thereby change it. This does not agree with the insects, who after all, have very subtle organs of smell and taste. That is why they keep away. Of course, the cultivated plant must itself be well looked after..." (quoted from *die Lebendige Erde*)

WIND PROTECTION There are, of course, all kinds of winds: warm, dry; warm, wet; cold, wet; and all sorts of other variations that nature plays on this theme, depending upon elevation and one's situation on the continent. Protection from winds can be provided by planting windbreaks and shelterbelts of trees and shrubs. These windbreaks, large and small, can help against a variety of problems — against wind erosion in dry, sandy areas; by catching and holding drifting snow; serving as living snow fences, and they also hold the snow back, and feed it gradually into the agricultural land beyond. In other regions, they are helpful against hot, dry winds which come along just when crops of grain and corn are still young, and cause what is known as *firing*. The plants are literally *burned up* at their tops. Other tender plants, like melons, are often *burned* by cold winds, and held back or killed in their early stages. In the garden, on the farm, in citrus groves, the Great Plains, and/or in the Corn Belt, trees and shrubs can be used by man to create a kind-of private, local climate.

One of the first "living fences" used to bound prairie farms was formed out of the Osage-orange *(Maclura pomifera)*, a relative of the mulberries, native to the prairie area. Many trees and shrubs have been used as farm windbreaks and shelterbelts, and proved to be adaptable in different areas, over a period of years.

WOAD WAXEN *(Genista tinctoria)*   *(See* LEGUMES*)*

WOOD ASHES   *(See* CELERIAC, POTASH*)*

WORMS, INTESTINAL Carrots are good to cure worms in goats. Tansy and mulberry leaves are curative of worms in horses, and sometimes also in cattle.

WORMWOOD *(Artemisia absinthium)* Gardeners may have noticed that plants growing close to Wormwood do not thrive. Experiments written up in 1940 show that Wormwood inhibits the growth of Fennel, Sage, and Caraway because of its toxic root excretions. Another scientist isolated a toxic substance named *absinthin*. Rain water flowing over the leaves, washes out this soluble, toxic absinthin, which soaks into the plant crown and into the surrounding soil, where it remains active for a period of time. Experiments were made, showing that Anise plants, close to Wormwood, grew only slightly, whereas Anise plants twice as far away grew more than seven times as tall.

Wormwood should not be grown close to any other plants in the garden because of this growth retarding effect, which is especially strong in years of heavy rainfall.

Neither *A. absinthium*, nor *A. vulgaris,* make good companions for medicinal herbs.

Wormwood will repel "black fleas" (black flea beetles). It will also repel moths and protect nearby cabbage plants against the cabbage-worm butterfly.

Wormwood tea sprayed on the ground in fall and spring will discourage slugs. A Wormwood-tea bath will chase fleas from dogs and cats. It can also be used as a spray in storerooms, where it is said to keep beetles and weevils away from stored grain.

The tea may be used on fruit trees and other plants to combat aphis. Do not use too often as a spray on tender vegetables, as it may retard their growth. Also, do not use too strong a solution.

Wormwood extract and an extract from the tomato plant will repel flies.   *(See* FENNEL, SAGE*)*

YARROW *(Achillea millefolium)* Yarrow increases the aromatic quality of all herbs. In a small proportion, as in a border, Yarrow helps most vegetables. Yarrow will grow in a narrow bed, as it does not mind being trampled.

Yarrow hay or Yarrow tea are good for sheep.

On rather dry pasture land, a high share of the stand is often occupied by yarrow, which likes to grow with perennial rye grass, *(Lolium perenne L.)*. A study carried out by G. Mahlke, Halle, 1953), showed that when grown together, Yarrow strikingly cuts down on the protein content in the grass. This can be explained by the competition for nitrogen of the two plants. Yarrow also considerably increases the fibre content in grass. This cannot be explained as a competition for nutrients. When the mixture of the two plants was analyzed, it was found that it yielded about the same amount of carbo-hydrates (N-free extract) as the grass. However, the mix-ture yielded 40% more protein (on account of the high protein content of Yarrow), than the grass alone. This explains why, when on dry land a pasture consisting mainly of these two species has long been known to grow feed of high, nutritional value for cows.

Yarrow is, in general, a good companion for medical herbs. *(See* COMPOST PLANTS*)*

Appendix ꙅ ꙅ ꙅ

CHROMATOGRAMS-CHROMATOGRAPHY

A specific chromatographic test is one way to find out whether a plant is helping or hindering its neighbouring plants in their growth.

Chromatography in general is now widely used in fields of research in science and industry concerned with separating complex compounds into their components, or in extracting a single element present in a complex mixture in minute quantity.

In the field of paper chromatography, a specific technique was worked out by the late Ehrenfried E. Pfeiffer, which shows in the resulting picture (chromatogram) the quality of a substance or living organism, in addition to its quantity. This method is unique in that it gives a synthesized, whole picture, not just a purely analytical separation into colored spots and zones. While in the latter method the form (not the size) of a separated spot or zone is of minor or no importance, in the Pfeiffer method it comes to the foreground and is just as, or even more, important and significant than the coloring in its different intensities and shades. Although remaining on exact scientific ground it also leads into the world of forms and colors, with their own qualitative and biological values.

This specific technique can be applied in many different research fields, especially those concerned with quality such as agriculture, nutrition, soils, composts, plants, etc. In recent years it has also been used to investigate the relationship between different plants as to their likes and dislikes. One can observe this kind of symbiosis in nature throughout the growing season; and one can discover it in a shorter time in the laboratory with the help of Pfeiffer's Sensitive Crystallization Method *(See page 25)*. The specific chromatographic test is a simpler method and has been worked out also with the idea of providing the farmer and gardener with a relatively simple test which he can even do himself, if he follows exact directions.

The technique, briefly described, is the following: circular filter paper discs (Whatman #1 or #4) of 15cm diameter are provided with a small hole in the center for insertion of a wick, rolled from the same kind of filter paper. The disc is laid on an open Petri dish for support. In the Petri dish stands a small crucible, containing a 0.5% silver nitrate solution, which will now climb up through the wick, dipping into it, and is allowed to spread over the disc until it reaches a certain pencil mark (4cm from the center). Then the disc is removed for drying.

Meanwhile, the solution to be tested has been prepared. For example, in testing a plant relationship: one gram of seed (ground carefully in a nut mill) is extracted in a flask (Erlenmeyer) together with 50 cc of a 0.1% solution of sodium hydroxide for four hours, with intermittent stirring and mixing, in the first hour. The resultant supernatant liquid is poured off into a small glass beaker and 5 cc measured into a clean crucible. The meanwhile dried, sensitized filter paper disc is provided with a new wick, through which the test solution now climbs up and is allowed to spread out until it reaches a second pencil mark (6 cm from the center). Then the disc is taken out, the wick removed, and the chromatogram put up for drying, and developing its colors and forms in diffused light.

In this way, chromatograms are made from the seeds (or leaves or roots), of the two plants to be tested, each separately. At the same time, a third chromatogram is made from a mixture, in certain proportions, of both plant parts extracted together. Each plant gives in the chromatogram a characteristic picture as to color and form. The resulting chromatogram of the mixed plant extract is the important one and will now reveal, whether the plants have a good or bad influence on one another. A good relationship will appear in the chromatogram as a harmonious picture, still revealing the characteristic forms of the plants seen in the separate chromas, blended well together. In a negative symbiosis the resulting picture will be disturbed, with distorted forms, or the characteristics of the one plant may be impaired or even completely overwhelmed by the other.

By varying the proportions in the combined plant extract one can also find out whether the one plant is helpful for another plant when both grow together in large quantities, or when there are only a few plants of one variety growing in or around a large field or plot of the other plant. There is open here a wide field for further application and research, as well as in the sphere of companion plants generally, using Pfeiffer's specific chromatographic technique.

[Erica Sabarth]

DANDELION *(See page 31.)* Dandelion greens are highly nutritious; they are rich in Vitamin A, iron, and mineral salts. The young leaves are served as salad or cooked like spinach. Medicinally they are good for torpid livers. The roots dried and roasted make a good coffee substitute. Because of its alleged diuretic qualities French cooks call the *dent-de-lion pissenlit.*

References ❧ ❧ ❧

Biodynamics (1940-2008), quarterly of the Biodynamic Farming and Gardening Association, Inc.

Bode, H.R.: *Planta 30, 567,* 1940.

Carson, Rachel: *Silent Spring,* Houghton Mifflin Company, Boston, 1962.

Caspari, Fritz: *Fruchtbarer Garten,* Wirtschaftsverlag H. Klug GmbH, Muenchen-Pasing, 1964.

Cocannouer, Joseph A.: *Weeds, Guardians of the Soil,* The Devin Adair Company, New York, 1950.

Coon, Nelson: *Using Wayside Plants,* Hearthside Press, Inc., New York, 1957.

de Kock, P.C., *Science, Vol. 121, 474.* 1955.

Georgia, Ada: *Manual of Weeds,* The Macmillan Company, New York.

Gregg, Richard B.: *Companion Plants,* Biodynamic Farming and Gardening Association, Inc., 1943.

Helgeson, E. A. & Kanzah, H. P.: *Phytotoxic Effects of Aqueous Extracts of Field Bindweed and Canada Thistle.* North Dakota Agricultural Experiment Station, *Bulletin 12 [3].* 1950.

Khalijah, R. A. & Lewis, L. N.: *New Natural Growth Promoting Substances in Young Citrus Fruit. Science, Vol. 142, 400.* 1963.

Koepf, H. & Selawry, A.: *Application of the Diagnostic Crystallization Method for the Investigation of Quality Food and Fodder. Biodynamics, Nos. 64,65,67. 1962/63.*

Krasil'nikov, N. A.: *Soil Microorganisms and Higher Plants,* National Science Foundation and Dept. of Agriculture. Israel Program for Scientific Translation, 1961.

Lebendige Erde, Journal of the Biodynamic Movement in Germany.

Lichtenstein, E. P., Morgan, E. G., & Mueller, C. H.: *Agricultural and Food Chemistry, Vol. 12, No. 2, 158-161. 1964.*

Linder, P. J., & Mitchell, J. W.: *Root Exudates. Agricultural Research,* January 3, 1954.

Lippert, Franz: *Vom Nutzen der Kraeuter im Landbau.* Forschungsring fuer Biologisch-Dynamische Wirtschaftsweise, Stuttgart.

Loudon, J. C., *An Encyclopedia of Gardening,* London, 1824.

McCalla, T. M. & Duley, F. L. *Science, Vol. 108, 163.* 1948.

McCalla, T. M.: (1959) *Nebraska Agricultural Experiment Station Bulletin 453, 1-31.* See *Advances in Agronomy, Vol. 13, 1961.*

McCalla, T. M.: (1960) *Soc. American Bacteriologists, Proc. 30,* See *Advances in Agronomy, Vol. 13, 1961.*

Mann, H. H. & Barnes, T. W.: *The Competition Between Barley and Certain Weeds Under Controlled Conditions. Ann. Appl. Biol. 34, 252.* 1947.

Notes and Correspondence (1950-1953), publication of the British Biodynamic Agricultural Association. Broome Farm, Clent, Worcs.

Percival, John: *Agricultural Botany, Theoretical and Practical.* Duckworth, London, 1936.

Pfeiffer, E.E. and Riese, E.: *Grow a Garden*, The Anthroposophic Press, Inc. New York, 1942.

Pfeiffer, E.E.: *The Earth's Face*, Faber & Faber, London, 1947.

Pfeiffer, E.E.: *Soil Fertility, Renewal & Preservation.* Faber & Faber, London, 1947.

Pfeiffer, E.: *Studium von Formkraeften an Kristallisationen.* Dornach/Schweiz, 1931.

Pfeiffer, E.E.: *Weeds and What They Tell*, Biodynamic Farming and Gardening Association, Inc., 1960.

Star and Furrow and Members Bulletin (1953-2008), publications of the British Biodynamic Agricultural Association.

Stout, Ruth, *Gardening without Work*, Devin-Adair, 1961.

Stout, Ruth, *How to Have a Green Thumb without an Aching Back*, Exposition Press, 1950.

U.S. Department of Agriculture: *Trees, the Yearbook of Agriculture*, 1949. U. S. Gov't Printing Office, Washington 25, D.C.

West, J. H.: *Growth Promotion and Control in Plants.* The Soil and Crop Science Society of Florida, Proc. 22, 44-48, 1962.

Whitfield, C.J.: *Stubble Mulching.* Agricultural Research, 6-7. 1964.

Zim, H. S. and Martin, A. C.: Trees, A Guide to Familiar American Trees. Golden Press, New York, 1956.

Index ❧ ❧ ❧